Christian Girl

Human Depravity

Christian Girl

Human Depravity

ISBN/EAN: 9783337371234

Printed in Europe, USA, Canada, Australia, Japan

Cover: Foto ©berggeist007 / pixelio.de

More available books at **www.hansebooks.com**

HUMAN DEPRAVITY;

OR

SIXTY-TWO YEARS OF MY LIFE,

SHOWING THE TRIALS AND PERSECUTIONS OF A ONCE HAPPY FAMILY, WITH WORDS OF EXHORTATION TO THE PUBLIC.

BY CHRISTIAN GIRL.

DECATUR, ILLINOIS:
1888.

PREFACE.

I respectfully dedicate this book to my wife and children, hoping thereby to reconcile them unto myself, and to again reunite the broken family circle, and to vindicate myself before the public. Much has been said, and much has been written, relative to my case by the so-call "Christians" of the church and faith I professed. Having caught the leaders in premeditated sin relative to church matters, and giving them to understand I would expose them to the world if they did not right their wrong, they expelled me from the church. Not being satisfied with expelling me, they endeavored to prove me to be an insane man, but could not, and their last resort, and most devilish act, was to break up my family, and to bring my wife and children in open court against me. It is often said that "murder will out," so in this case the true status of affairs is pretty well understood by all, but there are certain mysteries connected with it that I shall bring out, and fully expose those implicated. It is a good motto, to weigh well your undertakings. This my persecutors have not done. They were not aware of the magnitude of their undertaking, and have begged me not to bring disrespect upon them and upon the church. "Right, not might," shall prevail; let the keen sword of justice cut whom it may.

Christian Girl.

TABLE OF CONTENTS.

Chapter I,	1
My Boyhood Days,	5
Bishop David Frantz,	56
George Cripe,	72
A. Bingaman,	84
John Phillips and Brother,	92
Frederick Buckingham and Wife,	97
The Sisters,	106
T. Quickell,	118
Daniel Wagner,	122
Isaac Wagner,	129
Jacob Wagner,	133
Leonard Wagner,	136
M. M. Eshelman,	138
Jacob Deardorff,	142
William Heil,	153
Leonard Foutz,	162
Johnny Wise,	163
Jacob Ulery,	165
Daniel Venaman,	166
Solomon Shively,	177
Bishop Calvert,	183
Elder Miller,	189
Jacob Replogle,	195
William Girl and Others,	198
Simon Nickey,	201
Rev. G. W. Wilson,	207

TABLE OF CONTENTS.

A Recommendation,	213
A Letter,	214
Scandal Lane,	220
Pleading to Editors,	221
A Letter From My Daughter to Her Uncle,	224
Three Words of Strength,	227
Clippings,	228
Mr. Thomas,	231
Mr. Antrim and Wife,	234
Exhortation,	241
Elder Stoufer,	245
John Metzker,	257
Expense Account,	254
Brother Moore,	256
Last Law Suit,	259
Jacob Miller,	271
Eternity is Drawing Nigh,	275
Exhortation,	276
A Hint To All,	278
Exhortation,	280
Prayer,	282

HUMAN DEPRAVITY.

HUMAN DEPRAVITY;

OR

SIXTY-TWO YEARS OF MY LIFE.

CHAPTER I.

In preparing this volume I find many difficulties to be overcome, as I am not an historian or even a scholar, consequently a literary production with such barriers in the way will advance slowly, but the object of the publication is three fold:—First, I wish to again unite my family, and to inaugurate a mutual family love that should be all prevalent in every family. Second, To vindicate myself to the public, and to prove that what has been said by me is the truth; but being single handed in the combat, my evidence has been broken by false and malicious testimony, manufactured by the would-be leaders of the church. Third, To warn others, and, if possible, to deter them from becoming victims, and being preyed upon by a set of so-called "Christian believers."

While preparing the exposure of this particular sect and church, I deem it prudent to preface the volume with a brief sketch of my life, such as boyhood days, travels, and frontier experiences.

I was born in Stark county, Ohio, in the town of New Berlin, on the 15th day of July, 1825. My grand-parents were born in Germany, and emigrated to this country and settled in Pennsylvania, consequently I am of German descent. My mother was raised in the State of Maryland, and was a Palmer by name. I had five sisters and seven brothers. My parents called me Christian, nicknamed Chris. At or about the age of four, my parents sent me to a summer school about one mile distant from home. The school house was made of logs, and the seats of boards, ornamented here and there with curious jack-knife carvings. No doubt the main object of my being sent so young was to keep me out of mischief at home. My first teacher was a school miss. She died long ago. But her memory lives, and lingers bright as ever; and her image comes unbidden to my mind's eye whenever I think of my A, B, C's. Like all other mischievous boys, I sometimes came to grief by disobeying the dictates of the teacher. Upon one occasion I remember very well how the "school marm," in endeavoring to administer punishment, drew me upon her lap, and in my squirming to get away an opportunity presented itself, and I grabbed her little finger in my mouth. I bit and she pulled, the result being that she very nearly lost her little finger. Of course her loss would have been my gain, had I not chosen to spit it out. In the evening she called me to her and told me to bring her some salve the next day; but I failed to do as she requested, and consequently got a severe flogging.

It always pays to be obedient to our parents and to our teachers, to our government, and to all who are in authority. "Do unto others as you would have them do unto you." We should all try and live near the golden rule, and were it my privilege to live over the sixty-one years of my life, how different, and how much more could I have made of myself. By doing the right we influence some one else to do right, and this, in turn, has its influence on some one else. Then like the pebble when dropped into the calm water, the circle widens until it reaches the farthest limits of the shore. Without this regard for others, and the centralization of all in self, civilization is impossible, strife, contention, and discord would stalk abroad over our land. Church and state would dissolve, families be separated, and governmental powers made impossible. There is too much discord, back-biting, etc., in our churches. The outsider stands and looks into the many churches, but he is afraid to unite with any for fear of his being beaten and driven about by some skeptic preacher that doesn't practice what he preaches. He does it for various reasons. He is an intelligent, well-read man, but he preaches for fame, to be honored, not by God, but by men. He makes some of his hearers believe that he works for the mighty *God*, when it is for the mighty *dollar*. He is not ignorant of the plain truth of the word of God, " That the love of money is the root of all evil!" I shall not confine myself to the skeptic preacher, as I want to be honest and fair with my fellow men.

MY BOYHOOD DAYS.

Dear readers, it will not be necessary to go into details of telling you that I am an illiterate or unlearned man. You will soon find that in reading a short ways in this volume; but I want to show to the world how I have been persecuted, mostly by a number of professors of religion, and I hope you will bear with me while I try to unravel my life from the cradle to the present time, so I will proceed.

I was born in Stark county, Ohio, one mile north of New Berlin, on the 15th of July, 1825. At the time of my birth Ohio was a new State and very thinly populated, and in and around the section where I was born was very heavy timber, but the soil was good and productive, and its fame was known all over our country, and it grew very fast in population, the timber was readily cleared away by the new comers, and good farms inclosed and improved; and by the time I left my boyhood home it was a great farming country, and the timber became very scarce. I can look back and remember many happy days in my boyhood.

My father was of German descent, my grandfather having emigrated from Germany to the State of Pennsylvania, where father was born. My mother was a native of Maryland; both of my parents being robust and hardy. As long ago as I can remember mother used to work in the field, reaping wheat or pulling flax. Wheat was reaped with a hand sickle. Handful by handful the golden harvest was gathered, and many women gave their assistance in this work. Now, the self-binding harvester does the work of several hundred men with their sickles as in the days of yore, and the women confine themselves exclusively to indoor work.

The first laborious duty assigned to me was to carry water to the fields of wheat or flax, where, perhaps, one or two of my older sisters and mother were pulling flax.

When the flax was dry enough it was hauled to the barn, where the seed was threshed off, and the straw spread out on the meadow grass to bleach. When all this had been done, and when it had passed through the various stages of manufacture, we boys had the "pleasure of breaking in" a new shirt or a pair of new trousers. Do you fully understand the meaning of "breaking in," in this case?

We presume you do not, unless you have been the possessor of a linen suit; unless you were suddenly dropped into a thorn bush, or laid down among nettles, you would not fully grasp its meaning.

Well do I remember my mother's amiable disposition. She possessed an even temper; was mild, pleasant and cheerful; industrious, patient, and charitable.

Says Napoleon; "The future destiny of the child is always the work of the mother." We inherit from our mothers all those attributes which make us great, and owe our sudden downfall to none of her teachings. "A father may turn his back on his child," says Washington Irving, "brothers and sisters may become inveterate enemies, husbands may desert their wives, wives their husbands, but a mother's love endures through all; in good repute, in bad repute, in the face of the world's condemnation, a mother still loves on, and still hopes that her child may turn from his evil ways and repent. She sees her son, who has committed murder and is led forth for execution, still she remembers the infant's smiles that once filled her bosom with rapture, the merry laugh, the joyful shout of his childhood, the opening promise of his youth; and she never can be induced to believe him at all unworthy."

Even He, that died upon the Cross, in the last hour, in the unutterable agony of death, was mindful of his mother.

The mother's love is indeed the golden link that binds youth to old age. Through the memories of a mother's love I am able to recall many incidents of my childhood days, for she it was who made home dear to me.

In the language of that drunken tramp printer, Wordsworth, who wrote that deathless lyric poem, "The Old Oaken Bucket," in which he says:

> "How dear to my heart are the scenes of my childhood
> When fond recollection presents them to view,"

the author finds his sentiments, and many pleasant memories of the old home are revived and pictured in the mind after having lain in oblivion for a half century.

In my old age I find comfort in the thought of home, its surroundings, and how I enjoyed the company of my brothers and sisters. Home! no word in the English language approaches in sweetness the sound of this group of letters. Out of this grand syllable rush memories and emotions always noble.

The murderer in his cell, his heart black with crime, hears this word and breaks into grief and sobs. It brings to him the sweet memories of childhood when, perhaps, he was a good boy under his mother's care. He thinks of his mother, and of her kind words to him; of his youthful aspirations and future desires. What makes this word so touching to this rough man? Why, it was from home, the spot to which his heart is tied with unseen cords and tendrils,

When I was but four years old my parents sent me to summer school, in a log-school house. I was a mischievous lad, and one day my teacher took me to task and sat me on her lap. The scholars all laughed, and I squirmed like a captured rat. She proved too much for a four-year-old, of course, until I got hold of her finger with my teeth. She certainly got the worst of the affray, until the next day, when she gave me a sound flogging. My school days were few, for after I became old enough the duties of the farm devolved upon me. I was a gay and active youth, and took great interest in farm work, and especially in plowing. I found great pleasure in breaking colts to ride. Wheat raising was the principal farming, and father raised large crops of it; sometimes as much as one thousand bushels, and often as high as two thousand, all of which was tramped out with horses on the threshing floor; and at night we boys used to have a picnic in the barn, chaffing out the wheat that we had tramped out through the day. I used to have to ride day after day, and lead other horses in threshing out a crop. It was very tiresome work, and to clear up a crop of one thousand bushels of wheat required several weeks' hard labor. The immediate duties of the farm kept me from school the most of the time, and hence I only studied the three "R's,"—Readin', 'Ritin', 'Rithmetic.

I lived at home until 1846, when I was twenty-one years of age. I then decided to go to Indiana. I was accompanied by John Mishler and Mr. Provand, two young associates of my age. I had never been away from home very much, and when the day of leaving came I left the old home with tears in my eyes. I walked to Magadore, about five miles off, where I met my friends who were to accompany me. We went up into Michigan first, and afterward to Goshen, Elkhart County, Indiana, where John Provand, brother of my companion, lived. Here we were kindly entertained, and, with blistered feet — for we had been walking a great deal,— we found an excellent place to recruit.

After spending three years in the Hoosier State, and suffering with chills and fever for at least one-half of the time, I decided to go to California — the land of gold. Just at this time the gold fever was prevalent, and many men from all parts of the Union rushed to this far-off western country to seek their fortunes. At this time there were no railroads constructed through the western states and territories to the Pacific coast, and the journey must either be made overland across the plains, or by the way of the ocean, around Cape Horn. Either route was long and perilous, and many an ambitious man, full of hopes and future desires, lost his life in the attempt to reach the "promised land."

While in Indiana I had "a tough time," so to speak. Times were hard and money scarce, wages low, nine or ten dollars a month for a good farm hand. I chopped cord wood for twenty cents a cord. Many persons do not know the amount of hard labor in cutting a cord of wood. A good axe-man can chop two and a half cords a day by working very hard and keeping at it for twelve hours.

The winter of 1849 I went to school a few months to Mr. Pease, who talked of going to California. I became so enthusiastic over glowing accounts of the country and the immense fortunes that were being made, that I stopped going to school and began looking for an opportunity to go to California. I soon struck an old fox hunter who was going, and I made arrangements with him to make the trip. We talked over the plans and decided to start in March, 1850. I did not have money enough to pay my way. So he agreed to take me to the land of Eldorado, and I should work for him in the mines for two years; he was to board me, clothe me, and I was to give him one-half of my earnings. We wrote up an agreement to that effect. I had a young horse worth one hundred dollars, which I traded for a Canadian pony, which was to be a part of the team to carry us through.

On the first of March we rigged up our teams and set out for St. Louis, Mo., where we shipped our goods and stock to Leavenworth, Kansas. There we camped for three weeks, waiting for the grass to grow enough to support our stock across the plains.

When we were ready to start from there we joined others and formed a company and selected a captain, whose duty it was to go ahead and select suitable places for camping, and act in the capacity of "boss," as we called him. All the names of the company were enrolled, and every evening two men were selected to serve as guards for the fore part of the night, and two for the after part. We were in danger of the hostile Indians, and the guards were to fire a gun as a signal of danger, when all men should be in readiness for battle. About the third night after we left the settlements, we camped early, staked out our horses to graze, built a fire and all gathered around it, for it was chilly, and selected guards for the night. A young man from Illinois, whom many of us thought to be cowardly, was chosen. About eleven o'clock, while everything was still, and all of us were enjoying a "good snooze," we were aroused by a shot from a gun, and all of us sprang to our feet and seized our arms. On inquiry, we found that the supposed excitable "Sucker" had seen an Indian sneak up towards our horses, and had fired at him. Soon after one of the men noticed that one of the horses seemed uneasy, and he took a lantern to look after it. On examination we found that an old mare had been shot, which proved to be quite a severe flesh wound. Next morning we began to run the "Sucker," telling him he must not shoot our horses instead of Indians. We hooted and laughed at him, but he still claimed he had seen an Indian, and finally proved his assertions by showing moccasin tracks in the dust. It was supposed the Indian held something in his hand, which the bullet struck and was accidentally glanced to one side and struck the animal.

Next morning we pursued our journey over hill and dale to the westward, where we struck the Platte River; here we began to notice herds of buffalo. At night we camped on the bluffs about three miles from the river. Next morning, about daylight, we saw herd after herd of buffalo running from the river bottoms. We supposed that the grass had been set on fire by emigrants who had camped on the north side of the river, causing a stampede. I think there certainly must have ten thousand buffalo crossed the bottoms that day. We, concluding that it would be unsafe to travel, camped all day, and having noticed several buffalo calves falling behind the old ones, we thought they must be tired. I proposed to some of the men that we take horses, cut them off from the herd and capture them. We did so, capturing two about six weeks old. We brought them into camp, killed and dressed them. We found them nice and fat, and, as fresh meat was a scarce article, we had a feast on buffalo veal.

Next morning we journeyed over the prairies toward the land of gold. We saw quite a number of antelope. One of the company killed one, and we found it excellent eating. We became very much interested in finding game, and shot plenty of jack-rabbits, wolves and birds; and as we entered the rugged mountain region we saw occasionally a grizzly bear. We encountered much danger in crossing streams, as very few rivers had ferry boats. These mountain streams flow very rapidly, and are dangerous to cross. I remember very well what a time we had crossing the Green River, in Utah territory, which is a branch of the Colorado. This is quite a large river and has a strong current. We compelled our stock to swim, while we tightened our wagon boxes to ferry our goods over. We stretched a rope from shore to shore, which we used to help guide us across. It took us nearly all day to make the landing on the west shore, and then we continued our journey with renewed vigor.

Near the Sweet Water river, in Utah territory, we passed what is called Chimney Rock, a tall, natural pillar of stone, hundreds of feet high. Some of the emigrants climbed as high as they could to carve their names on this grand column.

For miles along the Sweet Water river were ledges of high rocks, and only one place could be found where we could reach the water to cross it, and this gap was known as the "Devil's Gate." Just after we left the river we began to ascend the Rocky Mountains.

We went up and down rough and rugged mountains, which it seemed impossible to do. When we descended the steepest places we tied a long rope to the hind axles of the wagon and locked all the wheels. With the rope the men held back, preventing the wagon from turning a somersault and crushing the horses. And thus it was that we went down many a steep slope of the Rockies. We reached the summit of the mountains about the middle of July, and saw it snow until the ground was white. We had gone right out of the scorching winds of the mid-summer sun into snow and ice, and it made us feel stupid. Some of our crew got what is called the mountain fever, and were very sick. This did not last long, for in a few hours we were down where it was warm again. Just after we got down off the Rocky Mountains we struck the Humboldt, which seemed to be quite a large river, but became smaller as we went down towards its mouth, until it sank in the sands. At its mouth is a large slough, or swamp, where the water sinks into the ground. This is a peculiar river in many respects. Its water is just the color of lye, and contains many parts of alkali. The water was hardly suitable to drink, but we had none other. It is the opinion of many that this river has an underground current which carries its water to the ocean.

When we reached the three roads leading to California I remember that the old fox hunter, with whom I was traveling, wanted to go to Fort Hall, about one hunded miles out of the way, to get a job of harvesting and buy some provisions. As I did not expect to stop on the way to do work, I did not like the idea of doing so, but the old hunter began to talk about the "articles of agreement," and that I should do whatever he said about the matter. I was not too hasty about the matter, and tried to reason with him. But no; he and his nephew would not listen to what I had to say. They knew it all. They had ignored the articles of agreement right from the beginning; they thought I was under obligations to them. Some of the best men in the company heard us talking it over, and urged me to take my pony out of the team and travel with them, as they were going a different road. I had decided to do so. As the old fox hunter had but one horse in fit condition to travel, he and his nephew soon saw their dreadful condition.

Being left with but one horse, away off from any assistance, was a serious thing, and demanded their immediate attention. The woman foresaw their condition and began to cry, which was too much for me. She had been very kind to me while I was sick, and I felt that it would not be right to take the best animal from their team, and leave them utterly helpless. Says I, "Don't fret over this matter; I won't leave you in such distress." I went to the old man and told him that I thought he had not treated me just as he ought, and further, I was ready to settle with him, as I intended to go the rest of the way with the other company.

I offered him my valuable pony for bringing me this far on the way. The old man protested against it, but finally agreed to take the pony and let me go.

I made arrangements with a couple of Missourians to carry me through from there. They were to board me, and I should work so long for them in the mines to pay them.

We started on the journey at once, by the way of Sutler's Cut-off through the Thousand Spring Valley. Here we found some wonderful springs; some hot enough to boil eggs, and others within a few steps of good cool water. This valley certainly contains many wonderful natural curiosities. We found many things to amuse and instruct us. How wonderful God has created this world, even beyond the comprehension of the wisest men.

Between the Humboldt river and the desert our horses became exhausted, and we had to stop to let them rest for a few days. When we came to the desert of forty miles, which lies between the Carson and Humboldt rivers, we had prepared for its crossing before we reached it by wading into the Humboldt river and pulling up wild clover, which we dried to pack on our horses. We decided to do most of our traveling by night, as it was scorching hot in the desert. We are about to mention a sad, but laughable circumstance which happened us as we were ready to start. We had placed a pack-saddle on the wildest horse we had, and loaded it with our cooking utensils, provisions, etc. When we got out into the desert about two miles the horse became frightened by the rattling of the camp kettles, got loose from us and ran like a buffalo, kicking, snorting, etc., until he lost all our stuff in the sand. What little provision we had was lost; it consisting of a half dozen biscuits and enough rice to make a rice soup or two. We gathered what we could of the rice and bread, for we were about starved out. We suffered most from heat and thirst, as it was terribly warm, and we had no water but that nasty alkali from the Humboldt, which we carried in canteens. We were making for the Carson river, and were very anxious to get there. Before we reached the river by five miles we found a good spring of water, where we stopped, watered our horses, boiled what rice we had left, and rested until the cool of the evening, when we moved on toward the river, where we found good water and good pastures for our stock.

We caught a few trout, and tried not to starve

out, though the chances were in favor of our doing so. Here we met some Government Relief Companies, who claimed to be sent out by the United States Government to aid those who were really in needy circumstances. They sold everything so high that few could afford to buy. Most emigrants were without money and provisions, and were compelled to get along half-starved.

Some would trade off their horses for provisions to those sharpers, who came out from California for that purpose. Many a horse was disposed of in this way for mere trifles, perhaps for twenty-five to thirty dollars. Could they have gotten through without this, and have let their animals recruited for a month or so, they could have realized two hundred dollars for them.

I here had some difficulty with the "Pukes" with whom I was traveling. We were almost starved out, and I could endure no longer. Those traders wanted a man to help them awhile until the throng of emigration ceased, then they were going to California. As they offered me four dollars a day and good boarding, I thought it a pretty good opportunity for one out of money and nearly starved, so I decided to accept their offer, provided I could satisfy the Missourians who had brought me several hundred miles. We had stopped a little ways off the emigrant road to let our horses graze. We were entirely out of provisions, and had nothing at all for dinner. While we were resting I made the Missourians a proposition to stop off the journey, and proposed to leave it to a committee of three what it was worth to bring me that distance. Then I should settle it by giving my note, and leave with them my clothes and other things as security.

They got mad, and one of the "Pukes" said "he would kick satisfaction out of me," and no sooner had he spoken than he ran up to where I was lying with my elbow on the ground, with my head resting on my hand, and kicked me just above the eye. I lay senseless for some time, and when I aroused again I found myself all bloody. I got up and staggered out to the road, where I saw some emigrants from Iowa. Here I related the circumstance to them, and told them just how it happened, and they seemed to sympathize with me. They gave me a good dinner, which I appreciated very much, being almost starved. They loaned me a double-barreled shotgun, which I was to carry through to Hangtown, California, and leave with certain gentlemen.

Here I was, away out in the wilderness, without a cent or a mouthful to eat. I could not think of starving, and made many vigorous efforts to shoot game of all kinds. I could occasionally shoot birds, which I roasted by placing them on a sharpened stick and holding them to the fire. I could not always find a bird, and so I went hungry most of the time.

I was in constant danger of the Indians, and came near being captured once. While traveling along the emigrant road once, I noticed an Indian trail which led across a valley that seemed to be much nearer by five or six miles. I thought I would go across, and perhaps I could find some game. I could see the covered wagons on the other side of the river. While I was passing along through a thicket I saw two Indians, about four hundred yards off, approaching me, walking as fast as they could, with their bows and arrows in hand. I kept a close eye on them and walked as fast as I could toward the stream. When I reached it I walked right into the water and soon found that it was deep in the middle, so I tossed my gun across to the other shore. I swam across and saw that the Indians had changed their direction. It was now late in the evening and I was hungry. I looked and looked for game of any kind, but could find nothing, so I had to be content to lie down on the ground to sleep without a bite of supper. I carried an overcoat which was my bed. I saw a great many emigrants, but it was of no use to ask assistance of any kind. Those who were liberal found many opportunities to give, and consequently soon became destitute themselves.

Starvation certainly looked me in the face. A young man and I went to a lake one Sunday morning and caught frogs, the hams of which we cooked, and found excellent eating; the trouble was that we could not get enough of them. I was always on the lookout for something to eat, as I had hard work to keep from starving. Once I discovered some half-grown ducklings about the middle of a small lake; I fired on them and killed two. Next thing was getting them; so I undressed, waded out into the lake, half tickled to death thinking what a feast was so near at hand. I reached them by wading in up to my chin. Just as I reached the ducks and had turned around to go back to the shore, I saw two Indians with my gun in their hands examining it. What shall I do? passed quickly through my mind.

There I stood out in the middle of the lake, and my clothes and gun in possession of the Indians. Horror stricken! What shall I do? My hair stood on end. Ah, reader, you cannot imagine one's feelings in such a plight! While I stood there for three or four minutes I seemed to think of everything. My mind wandered back to my dear old mother, whom I so dearly loved, and who always shared her son's sympathy.

I first thought I would swim across to the other side, but I at once saw that the mosquitos would eat me up. I became brave and started for the shore, and when I got about half way the Indians saw me and laid down the gun. While I was putting on my clothes they talked to each other, and pointing to a gingham necktie I wore, they said, "Swap! Swap!" offering me a small string of fish they had caught in the lake. I was glad to make the exchange, for I must confess that I did not enjoy their company in the least, and I skipped for the emigrant road, where I struck a couple of emigrants camping. I cooked my fish by their fire. I was very tired and lay down early to sleep. It became very cool towards morning, and as it was moonlight, I arose and plodded on my way.

There was great anxiety among the emigrants to reach the mining regions. They ran the risk of life to gain fame and fortune, and "to get a start in the world." Many a monstrous air castle was built on a frail foundation, and all, perhaps, demolished ere the "promised land" was reached. "For what shall it profit a man if he shall gain the whole world and lose his own soul."—Mark viii: 36.

As I traveled through the beautiful Carson River valley, with its numerous branches flowing down the sides of the mountains, I stopped by the wayside frequently to drink of the refreshing crystal water. It was in the month of August, and the snow-capped Sierra Nevada peaks were melting away. I was told at that time that the valley contained gold, and I met a few old "49ers" prospecting. I did not stop here, but pressed onward. With my gun on my shoulders, and an overcoat thrown across the gun, I ascended the rough and steep Sierra Nevadas. As I approached the summit I struck the snow line, where I found the snow deeper and deeper; at places it was forty or fifty feet deep.

When I reached the summit I was surprised to find one of my cousins from Indiana, who was traveling with a company of twelve, led by Captain Cary. My cousin invited me to travel with them the rest of the way, which I did, and found the whole company to be gentlemen. We were three and a half days in making the descent of eighty miles to Hangtown, California. Here we found a number of miners at work. Some were doing pretty well, while others were almost starving, and some offering to work for their board. As soon as we got there I began to inquire for work. I first spoke to the store-keeper. He said, "I can give you work if you can chop." I was a good chopper and knew I could please him. I stayed all night with him, and in the morning ground up an axe and he pointed to a tall pine tree about two feet in diameter. It was very warm, and as I was weak and tired out I could not stand it to chop very long without rest. When I had felled the tree I began to break off the foliage, and put it into a bag, which was to be used in the store as a carpet to keep down the dust, the floor being the bare ground. I was paid eight dollars for the two day's work, which I invested in provisions and started for the south fork of the American river, to look for work. There I found Captain Cary and his company, who had bought a claim, and, being inexperienced in mining, could not make it pay. They gave up their claim and disbanded the company.

My cousin and I went back to Hangtown to try the mine there, but could not make it pay. We heard of Mr. Lynch, of St. Louis, who was working a mine not far off, and who was paying seventy-five dollars a month for miners. We boarded ourselves and did our own washing and mending. At the end of the month my cousin thought he would not work any longer, so he demanded a settlement with me. He claimed that I owed the company eleven dollars for traveling with them about eighty miles. He said there were eleven of the company, and that they had talked the matter over and decided that eleven dollars was about right.

I paid him the money, but learned afterwards of Mr. Cary, captain of the company, that the company did not charge me a cent. My cousin had also urged me to take a pair of old pants and a pair of half-wornout boots, which I supposed he gave to me, but charged me double price for them. I paid his unreasonable demands. After he got back from California he told that he had saved me from starvation. That they were passing along the road and saw a man lying near by, and giving a kick turning him over, was surprised that it was his cousin Girl, and that he took me to camp and nursed me. These are untruths. A lie doesn't hurt, but the truth does sometimes. Our good Master says that a liar cannot enter the kingdom of Heaven.

I worked three months for Mr. Lynch, and in the meantime had became a good miner. I went to work on my own account, and made from four to five dollars a day. One expects to find something rich all the time, and, indeed, it is very exciting work, so much so that one is likely to work too hard. Notices in the newspapers of miners striking a rich find had a tendency to lend encouragement to the miner. What miner did not dream of fame and fortune! Yet how few ever found either?

Sunday was the main business day in the towns. The miners always went to town on Sunday to lay in a supply of provisions for the coming week. Sunday was also the great holiday and day of sport and gambling. I have seen piles of gold as large as a half bushel measure on the gambler's tables. The gamblers treated the miners liberally to get them drunk, so they could win their money. The poor fool miner would often lose his week's earnings on Sunday gambling with these sharks.

The territory was overrun with all nationalities, anxious to find fortunes. Chinese, Spaniards, French, Irish, Dutch, Africans, Hungarians, Danes, Swedes, Hollanders, etc., all found a place. The native Californians are the Root Digger Indians, which were numerous then. They had some curious habits, which should be mentioned here. They used a basket made of willow to cook their soup in. They first tightened the basket with pitch pine gum, and cooked their soup by heating stones and putting them into the soup. When the soup has cooked enough to put their hands into it, then they are ready to eat it, which they do by dipping it up in their hands.

In 1852 I left the mines on account of scarcity of water high up in the mountains. I went to Sacramento and San Francisco, and from there to the country, where I worked in harvest. After staying in the valley for three months and recruiting, Mr. Isaac Wamsley, from Southern Indiana, and I, bought us a Spanish pony and rode back to Mosquito Cañon mines. We worked together a short time, but our mine failed, and Wamsley went north. After a while I went into partnership with a Yankee, and a little Spaniard, who used to follow whaling on the sea. He was a native of Guann Island. In the winter of 1853 the weather was so wet that the roads became almost impassable for teams. It rained every day for six weeks; provisions became scarce in the mountains, and some of the miners killed venison and managed to keep from starving. We had a fair claim in shallow diggings, and we decided to buy a couple of pack-mules to carry provisions from the city. We did so, and I went to Sacramento to get provisions, which I bought at a wholesale store. While in the city the levees of the Sacramento river gave way and flooded the city. In many streets it was too deep to ford, and so I got my mules over and hired a boatman to row my provisions across so I could leave the town. It cost me just eleven dollars to get out of the city with four hundred pounds of provisions. Before I got home I encountered more difficulty. However, it did not amount to much. When I was within five miles of home I passed a little grocery store kept by an Irishman, who came out and claimed one of my mules. He said someone had stolen him out of his pasture. I told him that I bought the mule of an old gentleman who had come from Nevada. He said he knew better, that it was his mule and he wanted it right then. I then told him that I doubted very much whether he ever saw the mule, or whether he ever owned a mule. The Irishman got mad and began to pelt me with stones. I would not take that, so I got off and convinced him that two could work at that, and make it livelier, too. I soon sent him in a hurry to his business.

When I reached the mines I found the little Spaniard and the Yankee in good spirits, as they had struck plenty of gold.

Miners had hard times, and a great many were no better prepared on leaving the mines that when they came. "It is not all gold that glitters." Mining is to be compared with a lottery; uncertain business, and few become wealthy. The best mines were deep in the bowels of the earth, and it required capital to reach the precious metal.

Many miners spent their money freely, and especially so at the saloons. Twenty-five cents for a drink, and the same for a cigar. We had no silver or gold change then, and so the dust was weighed out on scales for that purpose.

Many a man who left a happy home in the East went to California and turned out to be a regular gambler, drunkard, and thief. Many murders, robberies, and other depredations were committed. I saw four men hanged at Coloma, Eldorado County. One was a married school-teacher, by the name of Crane, who had fallen in love with a young girl. She wanted to marry him, but her parents objected because he had a deserted wife and children in the East. He shot her.

Another man was a gambler, who had cut the throat of a fellow-gambler in a quarrel over a game of cards.

In 1856 I took a very sudden notion to leave the "Golden State," though I did so with some regrets, as California has a delightful climate and many other attractive characteristics. During the time I spent on the coast many improvements were made, and the State had doubled in population and changed from a territorial government to one of the Union. It was admitted as a State in 1850.

I purchased my ticket at San Francisco, via Panama, to New York. While waiting at San Francisco for the steamer, a ship arrived, and some of the passengers got into trouble on account of a drunken man, who went up to a fruit stand kept by the natives and asked for a quarter of a water melon; it was given him and he took a bite of it and threw it away and walked off without paying for it. This put the natives in fighting order, and they killed several innocent persons and wounded a number of passengers. The guilty drunkard escaped uninjured, but caused the death of several innocent people.

When we arrived at Panama the captain of the ship ordered that all passengers aboard go direct to the cars, which were to carry us across the Isthmus to Aspinwall, a distance of forty miles. He did so to prevent any trouble which might occur between us and the natives. It took about three hours to transfer us from the ship to the cars. It required but a short time to reach Aspinwall, where we waited half a day for the steamer which took us to New York.

A few days' sail from Aspinwall took us out of sight of land. We were in a storm which lasted twenty-four hours and was very severe; it looked as though we would be lost. The waves dashed high, and as they surged to and fro they seemed like great mountains. Many times they washed the decks clean, and struck as high as the sails. The sailors were compelled to tie themselves to the masts to keep from being washed away. While the worst was going on some were on their knees praying, while others were swearing. After the storm subsided we had a pleasant trip. The following language of Irving I find applicable in this case. He says in "The Voyage:"—"To one given to day-dreaming, and fond of losing himself in reveries, a sea voyage is full of subjects for meditation; but then they are the wonders of the deep and of the air, and tend to abstract the mind from worldly themes. I delighted to loll over the quarter railing or climb to the main-top on a calm day, and muse for hours together on the tranquil bosom of the summer's sea;—to gaze upon the piles of golden clouds just peering above the horizon; fancy them some fair realms, and people them with a creation of my own; to watch the gentle undulating billows, rolling their silvery volumes, as if to die away on these happy shores.

"There was a delicious sensation of mingled security and awe with which I looked down from my giddy height on the monsters of the deep at their uncouth gambols; shoals of porpoises tumbling about the bow of the ship; the grampus slowly heaving his huge form above the surface; or the ravenous shark, darting like a spectre through the blue waters. My imagination would conjure up all that I had heard or read of the watery world beneath me; of the finny herds that roam its fathomless valleys; of the shapeless monsters that lurk among the very foundations of the earth, and of the wild phantoms that swell the tales of fishermen and sailors."

How one lulls away the time in meditation as the ship glides through the noble waters of the deep.

At a distance we could see the whale, king of the sea, spouting water, and the huge albatross swaying to an fro, which enables it to fly day after day without stopping. We saw whole "flocks" of flying fish, which do not really have wings, but have peculiar side fins, which they use in making a leap out of the water, and sailing off a hundred yards or so. They have enemies in the air as well as in the sea. When the shark gets after them they fly out of the water, and they are in danger of the sea fowls.

We came by the way of Havana, Cuba, and were twenty-four days coming from San Francisco to New York.

I spent several days in New York, taking in the

sights, and from there I went by rail to Cleveland, Ohio, and from there to Canton, near my home.

Now, dear readers, I have written my history from my birth to 1883, and before Our Father in Heaven, I vouch for the truth of every word, and it is my heart's desire to place this book before this great American people so that every man, woman, and child that reads it may see how a once happy family was broken up by what are called religious teachers, but they are nothing but a set of imposters, and when their teachings were not in harmony with the word of God, they were too proud to acknowledge the truth and admit that they were wrong, but preferred to continue in their lying, in order to damage my character, because I would not partake of their devilish deeds, and in their last resort by their false swearing, they obtained a warrant for my arrest, and had me placed in prison on the plea of insanity, for the purpose of having me separated from them, so they could go on in their hellish crimes of sin unmolested; but Christ tells us in Matt. 5-3: "Blessed are the poor in spirit, for theirs is the kingdom in heaven."

Now I have returned to my native state, and in good health, hale and hearty. I looked around me and saw a wonderful change in the ten years of my absence. I visited the old farm where I spent all my boyhood days. The old brick house looks as natural as it did in years gone by. I went to the large bank barn, with its two threshing floors in it, where I used to ride the horses to tramp out clover seed, wheat and oats, when I was a boy. The barn seemed very natural to me — no change only a new roof. The farm was as good a one as could be found in Stark county, and there was a fortune in that farm for all of us. It was not in the rich and productive soil, but happened to be in the bowels of the earth, not in the many precious metals, like California's rugged hills and mountains, but in a bed or layer of coal in a meadow bottom, only from three to five feet from the surface of the earth. When I was a small boy I remember that we found coal and limestone in that bottom, but at that time timber was in our way, and timber was used for fuel instead of coal, but of late years coal has taken the lead in the way of fuel, and it became valuable, and there lay a fortune before our eyes, but we went away from it. My father came to the State of Illinois, I think in the year 1848, and moved to Stephenson county, where he died. He was the owner of about one thousand acres of the richest prairie lands, but they were not so valuable as the Ohio farm that he sold cheap because he got the Illinois fever, and so he left a certainty for an uncertainty, as I did in the gold mines of California. I bought an interest in a river claim on the south fork of the American river, lost money and time, left a certainty for an uncertainty, and worked for five dollars per day. After we left our shanty on the hillside, a lucky fellow struck a lead of gold that paid rich, that ran right through the shanty we had left to do better. I was told they found a lump of gold weighing several pounds, about four feet under the ground where we slept. Thus, you see, the mining life is very uncertain.

Well, I made a visit among my relatives and old acquaintances for several weeks, then took my leave from my two kind sisters and went to the State of Indiana, and made a stop of several weeks with my brother Joseph before I started for California in 1850, and when there with him I worked very hard, chopping cordwood and making rails.

When I left California in 1856 I came by way of Chicago on horseback. Chicago was, in that early day, a very small city, but gradually increased in population, until it now numbers in the hundreds of thousands, ranks third in commerce, and is almost king of cities in our great republic. From that city I made my way to Stephenson county, Illinois, where my parents live. I arrived at home after dark, and called as a stranger. One of my younger brothers came out, and I asked him if they could keep a stranger all night. Brother Peter answered that he would ask his father, who soon came out as lively as a cricket, and active as a young man, and said "yes, sir, you can stay;" and we had a very nice time when all of us got together. I remained in that neighborhood that winter, making my home with my brother-in-law the most of the time, and in the following spring, on April 26, 1857, I was married to Lucinda Brice Hart, of Stephenson county, and after my marriage moved on to my farm, and there remained until 1865, then moved to Macon county and bought a farm there, and have there resided ever since up to this date, living and enjoying a very happy and prosperous life.

I have lived, or tried to live, a moral life, and have tried to do my duty towards my fellow-men; but have finally concluded in my own mind, that according to the word of the Great Redeemer's commandments, there was something more demanded of us than just living a moral life, and believe there is only a step or two to make from a good moral life to becoming a good Christian.

I had, at this time, a great many calls, from preachers of various denominations, to join with the people of God and become a Christian. Myself and wife had been attending revival meetings that were held for weeks in what was called the Dunkard's brick meeting-house, north of me several miles. Here some very able speakers, such as preached the Word of Truth as I understand it, induced me to examine the Scriptures, for I had already had that passage of Scripture committed to memory: "Search the Scriptures, for in them ye think ye have eternal life, and they are they which testify of me."—*John* v-39.

Myself and wife had been attending those meetings very frequently, and my dear wife became converted to the Dunkard's faith and practice, and was baptized, I think, in January, 1872. It was in cold weather, and they had to remove the ice, about a foot thick, where I saw my side companion being led into the water with a number of other converts, who knelt down in the water and there received the grace of God. May God help her is my prayer. At that time I was converted and found my Saviour, and determined then and there to become a worker in the Lord's vineyard, and I meditated on those named verses, and tried to weigh well my undertaking. I was well aware that it was a big undertaking to live a religious life in this perverted and sinful world, but I resolved to go and work in the Lord's vineyard. I weighed my undertaking well. I knew that if I made a failure of my profession of religion that I would be laughed to scorn, and I resolved to try, and in a few days after my wife was baptized I followed her good example, and was baptized in the same stream, down on my knees, with the old pastor by my side, and there I made the good and solemn confession. I was also asked if I was willing to give and take counsel, to renounce Satan in all his pernicious ways, and whether I was willing to submit, and try to live up to those rules and order of the Church of God, which are laid down for all the members of the church, found in Matthew xviii, which I was taught by some of the well versed. The solemn scene made a great impression on my mind, and it put me to work to search the Scriptures. After we returned home I got out my testament, opened it, and I think the first chapter I turned to was Matthew xx., and read down to the seventh verse, when I went back and read the sixth verse over again. My wife's and my church relations ran very smooth for some three or four years, then came some trouble to us uncalled for. I will give a brief statement of the first troubles between myself and many of the members of the church. It was as follows: I had bought a quarter section of land of a preacher by the name of Leonard Blickenstaff, for which I had paid some cash, and given my note for $500, in yearly installments. The contract was for those notes to draw interest at the rate of six per cent. per annum. One note, that came to maturity at the end of four years from date, was made, by a mistake of the 'Squire, to draw ten instead of six per cent., and it was overlooked by me, and I expect by the other party; but to speak positively we will just say what we know of our own observations. I will speak for myself, that when that note at ten per cent. matured, it was placed in the hands of an executor, a Dunkard preacher by the name of Henry Troxil, who was appointed to settle up the estate of the deceased. He came to me for payment of the note at the end of four years, and it read ten per cent. interest from date. I told him it was a mistake, and asked him what I should do in the case. He gave me no counsel; he did not know; but assured me he was under duty to collect the note with the interest named. I went to see two of the deacons of the church, and asked them how I should proceed to get out of the matter. They said go and pay it to Troxil, and we will pay the four per cent. back to you. The note was paid in due time by me, without any words between myself and the executor, and when I demanded my four per cent. overpaid, they refused, and even denied the promises they had made to me. They were deacons in the church, and it finally came to a church trial.

Through the influence of the elder of that church, who was an uncle of the heirs, and by much persuasion, I was induced to place my case in the hands of a committee of three bishops who came from a distance, viz.: Henry Davy, from Ohio, Enoch Ely, from the north part of this State, and R. H. Miller, from Indiana. I proved before the said committee, by a brother in the church, that the deceased told him that he had sold his land to me at six per cent. interest on the notes given, but for some reason the committee decided in favor of the heirs, and threw me out of my just dues. I was not well satisfied with the decision, but had a timidness to say much about it; but some good thinking people advised me to submit to the case, with the admonition that it was better to give than to take. I have since had three or four church trials before that same official board of the church, but have not been able to get justice in said church.

I also had a trial over in the Okaw Church, before Bishop Wagner, but from the partiality for overreaching the line of our District Church, I was overpowered there, and my antagonist found in some way that the members of that church had nothing to ask of him, although he was guilty of helping to disturb the peace of my family. About one of the last church trials I had was with a man by the name of Quickell, who moved from our neighborhood to Pennsylvania. He wanted to get the advantage of me, and I reported him to Bishop John Raffensherger, of Clear Springs, York county, Pennsylvania, and the bishop took the case in hand. The agent whom Quickell had appointed to loan his money for him, found out that I was going to bring Quickell before the Council, and began to write untruths to him, in order to defeat me in my church trial, but I did not take much notice of such unfair proceedings, got an invitation to come down to attend the church trial, and gained a good victory among strange, imported brethren, and also got a good recommendation from the well-meaning old bishop, reading thus:—

"CLEAR SPRINGS, YORK CO., Nov. 4th, '83.

"*Dear Brother*—Your card of the 31st of October is at hand, and in reply I can say for myself and the brethren, as far as I heard from there, they spoke well of you, and I can, and the brethren say, your conduct was christianlike. You came here amongst us like a brother; you was willing to submit to the proposed plan in your's and Quickell's case, and everything in Quickell's case was settled up nicely and brotherly: but Quickell did not conduct himself to you as he ought, and is now divided from the agreement by the committee as decided. I have written to you a letter; I hope you have got it by this time.

J. H. RAFFENSHERGER.

The above is a true copy of the old bishop's card of recommendation about my conduct and transactions in the above case of a church trial between myself and a miserly man. A man that will agree to leave a matter in a committee's hands to be settled, and again to a committee, make up with his antagonist, and the next day go back on the committee's decision for the small sum of twenty-eight dollars, does well deserve the name of miser. "What does it profit a man to gain the whole world and lose his own soul?" as did Esau when he sold his birthright for a mess of pottage. Oh! that the people might see and feel that the love of money is the root of all evil. Oh! how we, as christian professing people, can yield to such low and debased things. May the power of God and all his angels descend down on such a set of hungry money-seeking, hypocritical set of Dunkards. They spurn pride by way of dress, and tell the wearers, if they chance to come to their meetings, that that hat, or that feather, will send them to hell unless they stop wearing them. Now, reader, look around you, and see whether you can see any pride in their works. I think you can see their fine barns, their fine horses, and their fine farms; then see their uneducated children in society; yes, there is ignorance. They will lie and steal from their members and everybody else to get this money to build up their pride, and then try to hide it behind their big-rimmed hats — and that is the reason they will not wear a fashionable hat, as there is not room enough to hide their sins behind it. You cannot rob your fellow man and go to heaven. No, hell is the place for such people. You are commanded by the Holy Bible to do the will of God, and you was told so by His Son, and now quit your thieving, and do His will and increase your faith.

I wish to say something more about the miserly one that I had some business with. Not being satisfidd with going back on an intelligent committee's work, he wrote to his Irish agent that wrote him the untruths in order to defeat me in the church trial, and the agent would not give up my papers until I paid him two dollars, he saying that it was my place to pay for the release of the mortgage he held to secure him on the loan of $1,800; but the Irish agent said he would hand it back to me as soon as he had time to inform himself, and find out if it was his place to pay for the release, but he refused to hand back the two dollars, and thinks he has played a cunning trick on me, when he is really doing himself more harm than to his opponent. I don't know whether the agent has kept the two dollars, or whether he sent it to the miserly brother. If I am allowed to judge, I would say that it is no more than likely they divided the amount between themselves; but the agent may have kept it all. He threatened to sue me for slander, and my wife hearing the rumor, went to see Mrs. Sowarn, wife of the Irish agent, and asked her if Mr. Girl slandered her. She said, "No, Mr. Girl did not slander me. Any one that said he slandered me tells a fib." That relieved my wife's mind, but it brought the agent into a tight place. His dear wife must have known that her husband went to court and swore that he would not believe me under oath. I will let the good thinking people judge who is the liar. The supreme ruler of the universe will judge all things properly. We will not forget that Paul had said the love of money is the root of all evil. I have heard from the old Bishop that the miser Quickell, with whom I had the trouble, got into a difficulty with one of his neighbors, by leaving his field open so that his neighbor's hog got into the field. He got after it with his dog and a pitchfork, and it was said that he killed the hog with the fork. The hog weighed about three hundred pounds. He afterwards went and told the owner that it was an accident; that he could not get his savage dog to let go the hog. We think the man was more to blame than the dog. He settled with the man for the hog, I think for ten dollars, perhaps one-third the value of it; but it seems that the owner of the hog was not satisfied after he found out that he had been told a falsehood. I don't know how it was settled in the end. It seems that dishonest men get out of one trouble only to become entangled in another. Will it pay to be honest? Every consistent man, woman and child will say, yes, it will pay. The author of the royal path of life says yes, and it is well said that "an honest man is the noblest work of God." The blessed Word of God says we shall provide things honestly before God and man, and live soberly and righteously in the present world.

Now, I have many things to say, in order to explain to the reader the number of enemies I had to fight with for many years, and a great number of persons were brought to court to testify that they would not believe me under oath, and they planned to injure my character, and have me sent to the Insane Asylum. We will name them out, and we will produce their evidence, and let the reader judge for himself. We will name a deacon in the Oakley Church. The deacon's evidence will be brought up in this particular case in order to show the whole conspiracy, and as I am sure it was all done through malice, we will endeavor to show, by their evidence, that it was one common bundle of malice by a certain clique, who joined themselves together to try and crush me in such a devilish way—to try and destroy me body and soul; and I will not hesitate in naming them, and in giving their evidence as far as I can remember it on the witness stand, as I think that every one who testified in that way has perhaps willingly perjured himself, and I want an equal chance with them. There is not one of them I would dare to trust in any case in court. The deacon's name I will now give. George Funk came to court and testified that he would not believe me under oath. I will produce his evidence to try to prove me insane, but he failed to make a single point, and exposed his ignorance before the court. He was asked whether he believed Mr. Girl to be insane, and he said, "Yes, he did," and his reasons for thinking so was that "Mr. Girl was riding around over the country and calling some of the people religious fools and hypocrites," and that Mr. Girl was abusing his family so that they could not live with him. He said two of his boys came to his house, and he even came over there to talk with the boys, and that he sent his son Chris. to drive him home, but said to the lawyer that they did not know what the old man was talking about to the boys, but he supposed the old man was quarreling with them; when it is an evident fact that I was reasoning, or trying to reason with my sons. On one occasion he went to see my son Charley, a minor, who has been under Funk's influence, and was employed by the Funks against his father's consent. After I was released out of prison and acquitted, and pronounced a sane man, and the jurors said Mr. Girl was too smart for them, George Funk contended that I was incompetent to judge, and of course things got pretty hot, and finally the matter came before the church, and he made his boast, "you bring me before the council, and I will show you that I have some friends there," for the very old bishop, J. Wagner, and the whole official board favored him, and they had no acknowledgement to ask of him, for he belonged to their precinct of church, though I had charged him with disturbing the peace of my family in various ways.

At one time I went on business to the deacon, and met him in his creamery. I tried to tell him how he was a tool in my antagonists' hands to help to destroy the peace of my once loving family. He got angry, and I walked away from him, and he after me with his fists clenched, red-hot, and bantered me for a fight. As we were both members of the same church, and were raised in one neighborhood in the State of Ohio, I thought it would be a shame for two old religious fools to fight a prize-fight. They intimated I was a fool because I stood alone fighting the devils, or, I might say, a legion of them behind the cloak of religion. The deacon made the remark, while he was attending court, that " the Dunkards don't fight very often, but when we do fight we fight like the teffle." He is like myself. He never learned the English language properly, and yet the above-named deacon went into court, and swore he had no malice against me.

I will give one or more evidence to the reader, then I will let them judge whether there was any malice in him or not. One of my witnesses testified that the deacon asked Mr. W. Kimberlin whether he thought that Mr. Girl was somewhat out of balance in his mind, or something to that effect. Kimberlin said to him that he never saw anything in Mr. Girl that he thought not right, and that he thought Mr. Girl a nice man. Funk then said: " Well, I told William Girl to go ahead and have him arrested, and we would help to pay his expenses in the case if he lost it." And the boy did arrest his father under such fellows' influence, and the pony purse was acknowledged by him and his son Chris. in the court room. His outlaw son, Chris. also swore on the witness stand that he had no malice against me. Well, we will see, as actions speak louder than words. I am a jocular man, and, of course, had a right to joke those ignoramuses. I told the old deacon and his son to look in the looking-glass if they wanted to see insanity. That got them up to fever heat, and the young outlaw sneaked up one evening after a Dunkard meeting was dismissed, and took me by the throat in the dark outside of the school-house, and demanded a promise to never throw such jokes at him, and to keep silent. I refused to make any such promise, and he struck me in the face. I had my overcoat hanging over my shoulders, and in backing up to get away from the young outlaw it fell to the ground. He was taken before 'Squire Lowry, and plead his own case, and swore that I pulled off my overcoat to fight him. I did not strike him at all. He was fined by the 'Squire, I think, eighteen dollars and costs for assault and battery. After that he made his boast that he had whipped the old devil once, and that he would like to lick him again, and at the same time said the old devil was crazy, and ought to be sent to Jacksonville. The young outlaw had a great deal to do with my family, and it was proved at court that he was a kind of a captain in what I called the Ridge mob, and he was the chief one among the sinners. May God have mercy on his poor soul. Without an extra effort the poor ignoramus will be lost. The poor sinner has come and quarreled with me, and called me an old liar and everything that was mean, in order to get me to fight, so that he might indulge in his criminal passions, not knowing that he was doing a low and debasing act; and when his old father was laid in his grave, his sister shed copious tears, for within her heart there was sympathy for mankind, but it was plain to be seen that he was a hardhearted wretch, for not a tear fell to moisten his eyes.

BISHOP DAVID FRANTZ.

This venerable old bishop has been an active worker in this case of persecution. He belongs to the Cerro Gordo cousining ring, one of the party that is in authority, and when he was on the witness stand, he testified as having dealings with Mr. Girl, and that he found him honest and fair in all of his dealings; and yet he came to court to help persecute me. I want the reader to read this bishop's evidence carefully, and see what a big deposit of religion he has. He is compared to gold, but is poor in precious metal, and abundant in alloy, and especially in cheap stuff.

As I want only an equal chance with my fellow-man in order to establish my rights, I have to be personal in this matter before us. I shall be so personal as to name out all that took an active part in this heathenish persecution to falsely imprison a man, and confuse a once lovely and pleasant family. I shall especially bear hard on those who have long been in the pulpit teaching others, marking out the road to heaven to their congregations, and claiming that God has ordained them to speak, and teach the people how to get to heaven, pretending to have a great zeal for God, and great care for the flocks that are under their care as a shepherd. Those pious-looking shepherds, so-called, would preach up the parable of our Saviour, that the good shepherd will leave the ninety and nine and seek that which is lost. I will single out those who made a pretense that they were carrying out the Saviour's instructions in those parables, as they claim that they have tried to carry out the Saviour's plain commands. As we shall know the tree by its fruit we will now give some of the fruit of those teachers of the Divine Word, by giving their actions, as they speak louder than words.

I clearly and distinctly remember the testimony of David Frantz at court, with the solemn oath he gave to God and to many witnesses, to testify to the truth and nothing but the truth. To that effect has this D. Frantz made his solemn promise before God and man. We will now give the evidence produced on the stand in the case of Lewis Beery, a son-in-law of mine, in the case of assault and battery. The old bishop was induced in some way by an evil spirit to be willing to come to the court to testify in behalf of the brutal son-in-law's case. Lawyer Bunn was my counselor. D. Frantz was called.

Ques.— Mr. Frantz, do you know Mr. Girl?

Ans.—Yes, I have known him fer years.

Ques.— How is Mr. Girl for truth and veracity?

Ans.— Not good.

Ques.— Mr. Frantz, would you believe Mr. Girl on oath?

Ans.— No, I would not.

Ques.— Did you ever have much dealing with Mr. Girl?

Ans.— Oh, yes; I had a right smart of dealing with him, and must say all the dealing I ever had with him was all fair and square, and he sometimes paid his obligations before they were due.

Ques.—Well, Mr. Frantz, I guess you don't like Mr. Girl very well?

Ans.— No, I wish we could run him out of the country.

Lawyer.—That will do, Mr. Frantz.

He left the stand like a shorn sheep. The Scripture says the followers of Christ shall all be of the same mind, speak the same things. This shepherd said he would like to run Mr. Girl out of the country. His colaborer said, in his own house, "Chris., I believe you are a christian; I believe you try to do right before God and man." These old veterans may deny these facts, but the truth will stand when heaven and earth shall pass away.

I will give some of the conversation that took place outside of the court-room a long time before I was arrested on false pretences. I went to both of these old bishops, just mentioned as colaborers in what is termed God's moral vineyard, and put the same question to both. I will give the exact language. I said to D. Frantz: "Did I honor you with double honor, as the Scripture demands that we should honor a bishop that rules well?" He answered, "I believe you did, Chris." I then said, "Don't you know I did well?" "I think you did."

I then approached his colaborer, J. Metziger, in the same way. He did not want to answer the question, and tried to draw my attention from it, but I told him to "hold on, Johnny; this is a fair question, and I must have an answer to it," and he gave it, "I believe you did."

After that they sent two deacons to hunt up a letter written by me to John Sowarn, and it was a very personal letter, not against the bishops, but against John Sowarn. It so happened that I was scarce of paper, and took a sheet to write to Sowarn that had their names on it. I had made an application to get a certificate of membership to go to Pennsylvania to attend a church trial. I will give the words of the certificate:

"This is to certify that C. Girl is a member of the Cerro Gordo German Baptist Church. J. METZIGER,
D. FRANTZ,
A. SNIDER."

This good old bishop has taken an active part in this heathenish persecution. He was willing to have a skeptic brother of mine brought from Indiana to testify against me the time the suit was withdrawn. He and Bingamer, and some more evil and false professors, went to the telegraph office with the blindfolder brother, George Girl, who was called from Muscatine, Iowa. I think George was persuaded by those old bishops, and an outlaw of a young preacher, to telegraph for the skeptic brother from Elkhart, Indiana. They were greedy to do evil, and that falsely and with evil intent. They did not like the idea of my son withdrawing the suit, but by some wise instructor the boy was induced to withdraw it, and those evil-doers telegraphed my brother in Indiana to come and assist them in their evil work to try and get me in the insane asylum. I wonder if the reader will fully understand these evil men's intentions that they had in their corrupt hearts. My brother at Elkhart refused to come. He had sense enough to stay out of such a scrape, and answered those ignoramuses that he would not come, and he afterwards wrote to one of my children that "Uncle George was a little insane instead of his brother Chris." Brother Joseph had that about right. Joseph is a deacon in the Dunkard Church, but is religiously blind. The old bishop, D. Frantz, has taken an active part in such unfair and ungodly actions, and we think him, and the devil-hired hand, A. Bingamer, the leaders of the mob. My wife told me that when they came to court to appear against me, Mr. Frantz and the boy preacher stood a little ways from her, and were holding a consultation about their defeat, as I had played a good joke on all my antagonists by disappearing the day before the trial. I made my way south to Mt. Vernon, away from those vicious devils, but they got together in the court-room and were wondering what had become of their victim, that they had tried to drive crazy, but were unable to accomplish their hellish design, and as they were defeated, it confused them, and they placed their evil heads together to plan out something else. My wife heard A. Bingamer say to Elder Frantz, "David, how would it be if we got Sister Girl to swear her life against Chris.?" Old Davy turned his eyes up so as to take a serious thought about the question just asked by a young outlaw, one that was hardly fit to live, and not prepared to die. The old bishop said, in answer to the young devil, "Well, I guess that wouldn't hardly do." My wife was weak-minded from the trouble of having her family all confused and broken up and scattered. My dear and well-wishing reader, what do you think of such movements from people that profess to love the Lord, and have the love of God in their hearts! Well might we say: "O ye generation of vipers." "Cease to do evil, and learn to do good," is the Saviour's instruction to his followers. I will let the reader judge in this matter, whether those pious-looking men have carried out the Saviour's instructions.

The Scriptures say we, as followers of Christ,

shall be all of the same mind; shall all speak the same thing. Well, we will see whether those old bishops do what they teach others to do. Old Johnny said: "Chris., I believe you are a christian; I believe you are trying to do right before God and man." I had a spy out to talk with old Davy Frantz, and he went up to the old man, and asked him whether he knew Chris. Girl. "Yes, I know the old devil." The spy then said to him, "I want to take a job of ditching from him; is he good for a contract that I would make with him, so that I would be sure to get my money." The old bishop said that he did not know, but repeated that the old devil was crazy. And still they asked an acknowledgement of that crazy C. Girl, and they are the ones that were guilty of disturbing the peace of my family. May God help him and his colaborers, to try and get out of their heathenish acts of devilment, in a more honorable way than to try and lie themselves out of such a plain case of ungodly work as they have been engaged in for many years. It looks as though they ought to make a new start in the divine life, and instead of their being teachers of the Word of Truth, let them be hearers of the Word of God and become converted, and let them repent of their sinful deeds, and we hope the good Lord will be merciful to them; but let them repent, and ask God for pardon. Let all good thinking people take warning, and provide things honestly before God and man, according to the Saviour's instructions, is my wish.

I will give Brother Frantz another push, and hope we can push him into the arms of our Saviour. He testified that they had me before the council concerning a personal letter that I had written to the Irish agent, and declared he would report me before our bishop. He did so, but the official board would take no action on it, but got the letter and brought it before the council, and had it read before the whole congregation, to expose me in my personality to the lying Irish agent, though the official board, nor the private members, knew the cause of my personal letter to the agent. What gospel authority did they have to have that letter read. It looks as though it was done to injure my reputation by an outsider, whom they had nothing to do with. Old Davy testified, that at the time that letter was read, or before it was read, the official board had met and counseled concerning Girl's case, and they would not take action on it, as it was brought before the church falsely. He said the official board came to the conclusion that Girl was not in his right mind, and they would not bother him. My dear and well wishing reader, I will tell you frankly and without any fear on my part, that that action was as false as the very devil, and the records of heaven will stand against such false proceedings.

Will the reader turn to the Psalmist, fifty-sixth chapter. I hope the old bishop will learn a lesson from this chapter. How can we believe such men that will give in such evidence. I will say with the Psalmist David:

1. Be merciful unto me, O God: for man would swallow me up; he fighting daily oppresseth me.

2. Mine enemies would daily swallow me up; for they be many that fight against me, O thou most High.

3. What time I am afraid, I will trust in thee.

4. In God I will praise his word, in God I have put my trust; I will not fear what flesh can do unto me.

5. Every day they wrest my words: all their thoughts are against me for evil.

6. They gather themselves together, they hide themselves, they mark my steps, when they wait for my soul.

7. Shall they escape by iniquity? in thine anger cast down the people, O God.

8. Thou tellest my wanderings; put thou my tears into thy bottle; are they not in thy book.

9. When I cry unto thee, then shall mine enemies turn back: this I know; for God is for me.

10. In God will I praise his word: in the Lord will I praise his word.

11. In God have I put my trust: I will not be afraid what man can do unto me.

12. Thy vows are upon me O God: I will render praises unto thee.

13. For thou hast delivered my soul from death: wilt not thou deliver my feet from falling, that I may walk before God in the light of the living?

Is it not so that God has sent them a delusion to believe a lie.

I will now give the actions of the third bishop, whose name is David Troxil, a colaborer with the two former I have named. I judge he thinks he ought to escape being exposed to the world. I have warned him often that he should save himself. He would never answer my letters. I have often invited him to come to my house and have a sociable talk, but he acted as though he was afraid of me. Perhaps he was afraid of catching the disease of insanity. A sister asked him, "Troxil, why don't you go and have a talk with Chris?" and he answered that he did not want to. He hung in with the other two, and one day I met him on the sidewalk and said to him, "Mr. Troxil, what do you think about this insanity business?" He said, "Well, Chris., I think you are not right in your mind."

Now the reader can see that the cousining ring were trained to all speak the same thing. It is an evident fact that they held counsel with each other to agree on that one thing;—to say that I was not right in my mind, and I am sure that they talked such nonsense to their children. They looked at me from their windows, and also in the place of meetings, as though I was a monster demon roving over the face of the earth.

I had written a number of letters to this old bald-headed bishop, pleading for my dear family, but could receive no answer to any of them, and when I spoke to him about them, he told me not to bother him any more, but I wrote some other letters, not only to him, but to the other bishops, who were his colaborers, and they would answer nothing. That induced me to write some very personal letters, and after Judge Smith granted them a rehearing, and set aside the judgment of five hundred dollars for false and malicious imprisonment, they then hunted up all the personal letters they could get, to bring to court to be read before the jury, trying to prove me insane because I wrote such personal letters. It took up about one whole day to read them all, as they had a big package of them. It was a good thing for the jury and spectators to laugh at. The court room was filled with people to listen to some of the most corrupt and devilish evidence ever heard, perhaps, in a court room. Those pious looking bishops tried to be very friendly around among the lawyers. I saw the two bishops stand and counsel together how to get out of the mischief they had got into in trying to break up a loving family. As they stood at one end of the court room, Lawyer Buckingham walked up to them two rich old shepherds and put his arms around their necks. I thought he was going to kiss them. He seemed to have a great love and respect for those pious-looking bishops; but I think, if the truth was known, and I think it is known to all good thinking people, that actions speak louder than words. So in this case. We know that the lawyers are all after money as well as the preachers, and all of the human family like money. The

Scripture tells us in plain words, to provide things honest before God and man. As the old Dutchman said to his son who moved out west, as he spoke the last words in their separating hour, "Well, son John, if you get out west you must try and make money. Try and make it honestly if you can, but if you can't make it honestly, make it any how." The counsel of the old Dutchman was good, if he had only left out the last few words. The author of the royal path of life says that an honest man's the noblest work of God. Has he got that right? Oh, yes; a child of ten years will say that is right. In addition to what has been said to those old bishops, I will quote some passages of Scripture for them to meditate upon. Let all carefully read the following:

1. Let as many servants as are under the yoke count their own masters worthy of all honour, that the name of God and his doctrine be not blasphemed.

4. He is proud, knowing nothing, but doting about questions and strifes of words, whereof cometh envy, strife, railings, evil surmisings,

5. Perverse disputings of men of corrupt minds, and destitute of the truth, supposing that gain is godliness: from such withdraw thyself.

Now turn with me to II Peter, ii: 1, 2, 3.

1. But there were false prophets also among the people, even as there shall be false teachers among you, who privily shall bring in damnable heresies, even denying the Lord that bought them, and bring upon themselves swift destruction.

2. And many shall follow their pernicious ways; by reason of whom the way of truth shall be evil spoken of.

3. And through covetousness shall they with feigned words make merchandise of you: whose judgment now of a long time lingereth not, and their damnation slumbereth not.

And I prayed for them as Paul did for Israel in his letter to the church at Rome, found in Romans, x:

1. Brethren, my heart's desire and prayer to God for Israel is, that they might be saved.

2. For I bear them record that they have a zeal of God, but not according to knowledge.

3. For they being ignorant of God's righteousness, and going about to establish their own righteousness, have not submitted themselves unto the righteousness of God.

4. For Christ is the end of the law for righteousness to everyone that believeth.

These passages of Scripture are written for just such old hypocrites. I will give a few more passages from God's word. Now please turn with me to the twenty-third chapter of Matthew:

1. Then spake Jesus to the multitude, and to his disciples,
2. Saying, The scribes and the Pharisees sit in Moses' seat:
4. For they bind heavy burdens and grevious to be borne, and lay them on men's shoulders; but they themselves will not move them with one of their fingers.
5. But all their works they do for to be seen of men: they make broad their phylacteries, and enlarge the borders of their "Hats."

It seems to me those old bishops are only hearers of the word, but not doers of the divine law. They go through a form of worship, but do not have the love of Christ in their hearts, therefore they sit perverted and confused.

GEORGE CRIPE.

I will now speak of George Cripe, who was one of the combination ring, as a fourth official, to persecute me. Before he moved to the town of Cerro Gordo, I met him at various times and told him of my grievances, and he seemed to sympathize with me in my troubles. At that time I was separated from the rest of my family, but lived with my wife in the town of Cerro Gordo. I had to make a sale of my personal property, and was compelled to quit farming and rent my farm to a stranger, on account of my boys and girls forsaking me, after I had proven myself a sane man, and showed myself competent to attend to all of my financial business; but the bad influence of evil men and women, and slanderous rumors, my family were overpowered by the enemy. They did not one come and speak peace to that family; and they all forsook their best friend. I hired a cook, had my youngest daughter to do the housework, while I took care of the summer crops. I invited my wife and children to come back home, but bad influences forbade them to return. I had to do the next best thing — made a sale of my personal property, rented the farm, traded the sale notes for a house and lot in the town Cerro Gordo, and moved neighbor to the old king of the cousining ring, Johnny Metzker. Some of my friends remarked that I was moving close to the old shepherd, and I answered that I liked to be close to my work, so as to get a good leverage on the enemy. I had counseled with my wife not to have anything to do with these neighbors, for they had an evil design in their hearts; but my counsel was disregarded, and the effect of it was an evil spirit was raised, and caused a separation between man and wife. Again there was lever power, but it worked against me.

I felt in need of a good counselor, and a cool-headed Christian man to heip to make peace in this confused family. I met Bishop Cripe in town, and asked him to bring his wife to our house and make us a visit, and he promised to do so. I told him that I was persecuted almost beyond endurance. He said he believed it, and I think he did; but alas! he never came to see us, but gave me a cold shoulder afterwards, and was impudent and saucy to me. I then commenced to write to him some mild letters, setting forth to him and others their duty as shepherds of the flock, and their duty as peacemakers. I quoted passages of Scripture, referring to a shepherd's duty as a peace-maker. I asked him to answer my letters, but never got a line from him,—nothing but a cold shoulder and a sour-looking face, and he became one of my worst and strongest enemies. It seems that he became a fool and a tool for his colaborers, and a strong pillar in my way. It made me shudder with fear to see the fourth big gun leveled to down me. I began to search the Scriptures afresh for encouragement, and it seemed as though I could hardly turn to any chapter in the New Testament or the Old, but what was full of encouragement for me, the persecuted one. For instance, the Psalmist, and the history of Job, and the Proverbs, chapter xxiv, "Be thou envious against evil men, neither desire to be with them. For their heart studieth destruction, and their lips talk of mischief. * * * He that deviseth to do evil shall be called a mischievous man."

I will give the actions of this pious bishop, and let the reader be the judge whether he is a mischief maker or not. We will show that he has taken an active part in helping to persecute me, not ignorantly, but willingly, with a heart full of malice against me, and I will try to make it plain to the reader. I fear no contradiction on the part of this big gun, who has stepped into the ranks of this cousining ring to help hide their sinful and heathenish act of separating man and wife, parents and children, and the reader will see that I have reached out to him for help to establish peace in my confused family. This I will prove to the reader, and without fear of contradiction from any of those four bishops. The truth will and must prevail in this ungodly, unfair, uncivilized and unchristian case.

This bishop was allowed to come to the Cerro Gordo Church and take a stand at the head of their council. He brought their business transactions before the church, with the three old bishops by his side. He is put at it to transact business for them, and he took the lead. I had often been to their council meetings, and nothing was said against my presence, until Cripe made a very ungentlemanly objection to me, even after they had commenced their transactions in a case between Replogle and Dr. Sayler, an appeal case that I had after I was illegally expelled from that church. I walked into the meeting house, took a seat about the middle of the room, sat by myself and listened to the work they had before them for about half an hour, and it seemed that they were not getting along with the case before them as well as they wished on account of the absence of some witnesses on both sides, so they came to a deep study how to settle the case before them. All at once Cripe pointed his long finger at me, and said, "There is a man who must go out of this house. He says the foundations of this church are rotten." He took me by surprise, as I had not said a word until I was almost forced to by the bishop's personal remarks. It appeared that this unfair and unprincipled matter was all planned out to give me a send-off out of the house. It seemed to me that it was planned by that clique of evil-doers, that if I did not go at the rough warning, they would put me out. A few of the officials raised to their feet and showed a willingness to carry or lead me out, and I thought it rather a rough assertion to make on a persecuted one. As it looked to me that they were making ready to take me out by force, I told them, and at the same time pointed to the cane lying on the seat next to me, that they had the will but not the pluck to do the dirty act of boosting me out of doors. I kept my seat, and one of the unreasonable bishops proposed that one of the deacons should go and get an officer to take me out of the house. The deacon started to go, and went as far as the door, when the bishop told him to hold on and come back. He then suggested that they adjourn, and have it said that they had to postpone the council on account of C. Girl. I told them to go on with their council; that I came there merely to hear the appeal case. I had not a word to say in the matter, but they were determined, and did adjourn. Cripe named a hymn to sing. I walked to the end of the bench, got a book, turned to the named hymn, and helped him to sing it through. Cripe then made a long prayer, and it was rather personal against me. I put in a few amens, when he prayed for God to correct the ungodly and the sinner, and it seemed that aggravated their case that was so falsely in their treacherous hearts. After prayer the meeting was dismissed, and I walked peaceably out of the house. Soon after that they put an article in the Cerro Gordo paper that they thought they would bring C. Girl before the justice for disturbing the peace of their meeting, but made the statement that they were a non-resisting people, and that they would let the irresponsible C. Girl go; but shortly after sued me before 'Squire Middleton, of Cerro Gordo, for disturbing a religious assembly, and got old 'Squire Barnwell, a big, fat, and lazy Methodist, to pettifog their case. Well, he tried the very best he knew how to convict me with the untrue evidence, to get a fine on me, but failed, and they had the costs to pay, and they got beat at their own game. Bishop Cripe and Eli Cripe, the blacksmith, were the prosecuting witnesses, and a pony purse was made up to pay their costs. They were noted for getting up pony purses, and it was proved on them that they had made up a purse for William H. Girl, when he had his father arrested on a plea of insanity. They went from bad to worse, with their heads full of malice, like a set of Judases who persecuted our Saviour.

I heard some rumors, through some of my friends, that they had reported me before the grand jury at Monticello, Piatt county, for the same thing. The jury found a bill against me, and the constable came and read his warrant to me, and I had to at once get a bondsman for my appearance at the next term of court. I tried to get the banker, John S. Coons, of Cerro Gordo, to go on a bond with me for $200, but he refused. I asked a few others, but they all refused, and I went to the constable and begged him to let me go to Decatur, the county seat of Macon county, where I lived. I gave him a note of $100 on my tenant for my appearance next day to meet him at Cerro Gordo, or go to jail to await my trial, or get some one on the bond for my appearance at court the following term, there to answer for the deeds done in the body. I came to Decatur and told my wife that I had to go and meet Constable Bell and go to Monticello with him, and that perhaps I had to go to jail there and await my trial. She shed tears and said that she was going with me. I told her all right. The next day we took the train, met Mr. Bell at Cerro Gordo, and went with him to Monticello. The sheriff there asked me what I was going to do about the matter. I said I had tried to get bail, and could not, and that I supposed I should have to go to jail. He said "No, you shall not go to jail," and proposed that my wife should go on my bond, and we fixed it in that way, and came home on the returning train. On the day of trial, or rather at the commencement of the term of court, I took lawyers Bunn and Nelson to attend to the case, as they had plead my case before 'Squire Middleton. As soon as court opened they made a plea to the judge, and told him that they had handled the case before 'Squire Middleton, and that Girl did not disturb a religious assembly, and the judge dismissed the suit, saying they had failed to state in what way Mr. Girl had disturbed them. Bishop Cripe and others were standing and sneaking around like a set of hungry prairie wolves, waiting to get a chance to give in bad evidence against me. They sneaked off as though they were shot at. One would now think they would stop their heathenish work, but they did not, and if the reader will bear with me, we will look at another of their evil intentions.

I met the saint of a bishop on the public road going from my farm to Decatur. I hailed him. He had his wife along. I was alone. He had a spring wagon. He stopped, and the following conversation took place. Says I to Cripe, "What business had you to order me out of the meeting-house?" He said; "They told me to order you out."

"Who told you," I asked, but he would not say.

"Did old Johnny and old Davy tell you?" but he would not answer.

I told him it was them; that they were too cowardly to order me out, and they got you to do it. He did not deny it. I then told him that if these bishops told him to cut off my head, he would have to do it, and after a few more words he jumped out of his wagon and bantered me to fight. I told him a few truths, and he went back into his conveyance, but things got pretty hot again by pouring the truth at him, and he jumped out the second time, and bantered me again to fight. I told him I was going to whip him with the sword of the Spirit, and he got back into his buggy and drove off. I was determined to get that man out of the way, as he was a tool and a fool for his colaborers. Some time after that racket we met in Decatur on the platform of the depot. He and another man were walking by me to get on the cars to go to Cerro Gordo. As he walked by me I said, "Here goes the Dunkard fighting cock." He colored up in the face, but passed on and took a seat in the last car. I walked around on the side where he was sitting, and told him he was a tool for those evil men, and that he had better come out on the side of right and do his duty. He made no reply, but got up from his seat, and said he was going to hunt up a policeman. As he walked out of the car, I slapped my fist, and said, "If you want to fight, now is your time. Come right here, I am ready for you. Come on, you cowardly fighting cock." He walked up in front of the depot where stood a policeman. I hallowed to him, and said I would help him to find one. I went half way, and sat down on one of the trucks. The bishop talked to the policeman and pointed towards me. He knew that I was being roughly handled by these evil and false bishops, and made no attempt to arrest me. The cars being about to start he walked past me, and I said, "Ha! ha! the police and sheriffs are my best friends; you old fighting cock, you are a cowardly man." I am aware that I did wrong, and wrote a card to the bishops acknowledging the same. "Blessed are the peace-makers, for they shall be called the children of God." I hope these old bishops will take a square look at themselves, and see themselves fairly.

We will quote some Scripture for their benefit. Let them read I Timothy, iv–1-2. Please go with me to the second Epistle of Paul to the Thessalonians, chapter second:

1. Now we beseech you, brethren, by the coming of our Lord Jesus Christ, and by our gathering together unto him,

2. That ye be not soon shaken in mind, or be troubled, neither by spirit, nor by word, nor by letter as from us, as that the day of Christ is at hand.

3. Let no man deceive you by any means: for that day shall not come, except there come a falling away first, and that man of sin be revealed, the son of perdition.

9. Even him, whose coming is after the working of Satan with all power and signs and lying wonders,

10. And with all deceivableness of unrighteousness in them that perish; because they recived not the love of the truth, that they might be saved.

11. And for this cause God shall send them strong delusion, that they should believe a lie;

12. That they all might be damned who believe not the truth, but had pleasure in unrighteousness.

17. Comfort your hearts, and establish you in every good word and work.

A. BINGAMAN.

I will now proceed to give the evidence and conduct of the young preacher, A. Bingaman. He goes by the name of the devil's hired hand. He is not the Adam that was put into the Garden of Eden with Eve to obey God's commandment, not to touch the forbidden fruit. We think he dare not tell his wife that she has given him of the fruit, and did eat thereof. We have reason to think that his wife is in darkness concerning her husband's many false and treacherous evil doings to help throw discord in a once loving family. He was very active in taking pains to confuse my family, and he played an active part in being obedient to the elder brethren. I saw he was stimulated by the elders to take their counsel and obey their instructions. The scriptures say the younger shall serve the elders, but only in what is godly and right; but we will show the reader that that young preacher was rightly named by me, when I called him the devils' hired hand. We will prove to the reader that our assertions were true. He was willing to go with two of the old bishops, Johnny and Davy, as the third person to settle a difficulty between two members of the church, according to Matthew, 18th chapter, 16th verse:

"But if he will not hear thee, then take with thee one or two more, that in the mouth of two or three witnesses every word may be established."

The reader will see that I was the complainant, and that I was the one that was to take one or two with me, according to the Saviour's instructions, but the two old bishops stepped out of their way to bring the devils' hired hand along. He was obedient to the bishops' wishes, and went with them, and right there they all three violated God's law, not ignorantly, but willfully, and for a vicious purpose. They appeared before me, and introduced the third person that they thought they would bring along. I asked them what gospel authority they had, and referred them to the 18th chapter of Matthew, to the Saviour's instructions. I sent him away in a hurry, and gave the two elders a sharp reproof for their illegal proceedings and strange conduct. The young preacher was ever ready to obey those old bishops, right or wrong. He was the man to keep my family under excitement, to get up rumors of a slanderous character against me. My oldest son, with a heart full of malice, was urged on by him, C. Funk and others. When I was arrested the first time, that young preacher made himself very prominent as a counselor in my family, and especially with the oldest son, urging him on to make the arrest, but waited to be excused from being a witness in the case, which created some hard feelings between the conspirator and my son, who said that Bingaman had to go to court and testify. The suit was withdrawn after I came back from Iowa, by persuasion from myself and some of the good citizens. When Bingaman found out that the suit was to be withdrawn, he made it his business to go to the boy and beg of him not to withdraw the suit. I reasoned with the boy, and told him that I would prove myself a sane man, and in spite of all they could do, the suit was withdrawn; but the malice in those men and women still existed and the second arrest was made and I was lodged in jail. Then Satan had a chance to come in with his false evidence, and now for the devils' hired hand's evidence to prove me an insane man.

As he was put on the stand, he seemed to be glad that his turn came to give in his evidence. He was asked by Lawyer Bunn what his age was and where he lived, and then interrogated concerning Girl's sanity, whether he believed he was insane or not. He was very prompt and active, and seemed as though he was real proud in making answers to the questions. The lawyer gave him plenty of string to hang himself with, which he did in great shape. The lawyer said: "Mr. Bingaman, what made you think that Mr. Girl was insane?"

"Well, he wrote all over the neighborhood and got it in an uproar, and he called the preachers hypocrites and devils' hired hands."

Lawyer—"How far do you live from Mr. Girl's farm?"

Ans.—"About one mile."

Ques.—"Was you afraid of Mr. Girl?"

Ans.—"Yes, sir; I didn't go to my barn after night for fourteen weeks, on account of him, with the exception of one night. I had a sick horse, and I had to go out."

Ques.—"Mr. Bingaman, how did you manage that night?"

Ans.—"I took a body guard with me."

He did not say what that was, whether men or a pistol, or what. Soon after that he had my brother come to testify against me. He brought him right before my house to try and confuse my family. As I came out of the field I saw them both standing there, and I had to order them away. This goes to prove that he was not afraid of me at all, and that he swore to a lie. He also said that Girl got so ornery that they had to expel him from the church—and that was a lie. He also said that he (Girl) had tried to correct those old bishops, that may be true, but they would not be corrected, to the sorrow of all good christian people that love the Lord. If Bingaman ever was afraid of me, it was his own guilty conscience that made him afraid. It looks as though such men have no conscience. If they had anything like a good conscience they could not perform such heathenish and hellish persecution, willfully and maliciously, with their hearts full of malice, and still come to court and swear they had no malice against me. All good thinking people know better, especially those who heard their testimony. After I was acquitted by an honorable jury, some of the jury said that "Mr. Girl has got more sense than any that appeared against him," when this young impudent devil's hired hand remarked "that Girl has plenty of money, and he bribed the jury;" and he said to a spy that "if they would give him two hundred dollars he would put Girl where the dogs would not bite him." Would it be possible that such a combined set of fools would be able to send a sane man to an insane asylum. I have said to them that I would use their meeting house for an insane asylum for that cousining ring, and that I would be their superintendent and do their preaching for them.

I will have something more to say about the young deacon, Brother William Bingaman, as he has taken so active a part in screening those evil doers from the effects of their mischief.

He told me that he saw letters that were written to some of those old bishops that were not fit to be seen or read, that had no name signed to them. I asked him if he saw the letters, and he said he did, and read them, but would not tell me who they were directed to, but he kept telling me, you know. I told him he wrote those letters to the bishops, and that I signed my name to the letters I wrote. I tried hard to get him to say who the letters were directed to, but he would not tell, and finally I told him, before his wife, that I believed his brother had written them, and also believed that he had a hand in the work, because it would compare well with with their other hellish deeds, and the evil thoughts in their treacherous hearts.

If he was innocent about those letters, why could he not answer my fair questions. Actions speak louder than words. We will give some more of his actions.

He was brought on the witness stand against me, and there testified that I told him that I was sent by God to correct those Dunkards. I told him in the presence of his wife that he had sworn to a lie, and if a man lies to injure me, I will call him a liar if he is as big as an oak tree.

When on the witness stand he got very important and saucy. We thought for a while that he was going to run away with the judge, jury and lawyers, but they gave him plenty of rope to hang himself, and his evidence hung him in close connection with that cousining ring.

We suppose that his wife does not know that her husband has given in such unfair evidence against me. We have reason to believe that she is a zealous christian woman. I had a good many conversations with her and him, and we have talked about the evil of bad ruling in the church, and we talked about the evil influence of the cousining ring, and bad proceedings in the church, and always agreed on the evil we could see in the unfair ruling, when, all at once, W. Bingaman turned his coat. He would still try to deceive me in meeting me in a friendly way, declaring that he had nothing against me. I was determined to make him show his colors as I came with him from town, and I pressed him, when he flew off the handle and called me a liar. He got over the fence first, and then abused me. I told him I had to press him to make him show his colors; told him that he was a snake in the grass, and after that he was willing to openly stand against me, and even to testify to the untruth. His wife, I think, does not know of his unchristian principles, for if she did, she would sharply reprove him for such ungodly conduct; but as we know that this christian religion is a personal matter, she will not have to answer for his sinfulness. We read, "Let no man deceive you with vain words, for because of these things cometh the wrath of God upon the children of disobedience."

Please turn to Ephesians, chapter v, 15th, 16th, and 17th verses, and Christ will give the light.

"See, then that ye walk circumspectly, not as fools, but as wise, redeeming the time, because the days are evil." May God help those evil doers that have conspired together to do evil. May they take the admonitions of the 17th verse of the above-named chapter: "Wherefore be ye not unwise, but understanding what the will of the Lord is."

JOHN PHILLIPS AND BROTHER.

I wish now to expose the weakness of human nature of man, and show how we may be influenced on the side of wrong. I will now speak of two of my neighbors, John Phillips and wife and Hiram Phillips. John Phillips and wife belong to the Dunkards, and we have been neighboring together for many years. I have laid my grievances before them, explained to them all the unfair dealings of the official board of the church — how they have gulled me with false actions taken in that cousining ring — and they seemed to sympathize with me, and seemed to enjoy my reasonings on moral and Christian principles. Many a time in the long winter nights have we sat together and talked about the evil of the way in which the churches were ruled by a few individuals, and appeared to enjoy my company. They most always gave me credit for advocating the right side of moral and Christian principles up to the time of the appeal in the insanity case. I was overpowered by the false evidence of the official board of the church and others, and some, even from other churches, were induced to bring in false evidence in every possible way in order to beat me, and throw the costs on the persecuted one. They succeeded in their hellish undertaking, and I had to double mortgage my farm and other property in order to meet those unjust court expenses. I have paid all those unjust debts with a strong resolution. I have never allowed any sheriff's sale with the exception of a part of one hundred and seventy dollars that I failed to meet. Eighty acres of land were sold to secure that amount of costs, which I expect to redeem before long.

The suit was given against me by the false evidence, with a treacherous lawyer at the head of whole mischief, in order to make a little money out of the heathenish affair. Both of the Phillips' evidence was good and fair in my favor in the insanity case. They testified that they had conversed with me on various subjects at different times and places; that I had managed my farm as well as the ordinary farmers in that vicinity, and their testimony was all in my favor; but as soon as I got beat at court they turned their coats and declared that I was persecuting those old preachers. They thought that I was now overpowered, and that I must now go down to poverty and despair; that I had many enemies, and they wished to stand on the winning side. They were now willing to stand up with my enemies, and even helped, with their influence and arguments, to throw more grief on me. They said I slandered and abused those pious-looking bishops, and one day I was tackled by Mr. John Phillips on scriptural points. We were talking on various subjects, good and evil, and about the division in the Dunkard church. At that time a dividing element had sprung up in the church, some pulling off from the main church. Their excuse was bad ruling. I had many arguments with them on the subject of the evil of a separation in the church. Those who called themselves " old orderists," wanted to pull off from the main church. I told them that, united we stand, divided we will fall, but they would pull off, and it is now a separated church. Since I was expelled illegally from the church the old orderists have invited me to go with them. Says I, " What, to jump out of the frying pan into the fire? You are drones; you won't work, but pull off and find fault; we will call them backsliders and backbiters." Mrs. John Phillips knew my opinion about old orderism, but in an argument with her she said, in a sneering and insulting tone, " Well, Mr. Chris., you had better go and join the 'old orderists.'" It was an important assertion. She said it to aggravate me. I had asked that woman often to go and speak peace to my confused family, but she took her own way for it, and said she pitied William Girl the most of any of the family. I made it plain to both of them that the boy was persuaded against his better judgment by such evil men as A. Bingaman, D. Funk, and others. I, at one time, had confidence in John Phillips that he would stand by the persecuted one. I told him that I had a plan to make that boy come out with his secrets.

He wanted to know what my plans were. I told him to mob the boy and make him tell the secrets he held for such devilish mean folks. John turned against me, and went out and told the boy, and I was soon arrested again. Thus, instead of trying to make peace in the family, he was willing to take part with the evil man and woman. Thus, we can see what bad influences will do. It would have been a blessing for the boy if he had been mobbed, and made to tell the secrets he was induced to withhold against his father with a heart full of malice. It does not seem possible that a boy that was well raised could be made to try and break his father up against his own interest. I think Phillips put him up to arrest his father the second time. Thus, instead of making peace, they created strife and confusion. The Scriptures say, "Blessed are the peace makers, for they shall be called the children of God."

Now, my kind Christian-professing people, you had better turn your attention heavenward and put your trust in God, and not in man, as you have in time past, for man is a weaker vessel than your Supreme Maker. Man is liable to fade like the summer rose before the autumn breeze, and then all your hopes and fond aspirations are gone forever, but God will never fail you.

FREDERICK BUCKINGHAM AND WIFE.

It is my aim to expose all who have been induced by bad counsel to go into court and testify that they would not believe me under oath; but I will make some allowance for ignorance in the case of Frederick Buckingham and his family, whom I have neighbored with for many years. They moved from Ohio to Illinois, and they were very poor and much dependent on their neighbors in the way of farming implements. They would come and borrow plows, drags, wagons, and all tools that were used on the farm. They would also borrow flour, and flour sacks to go to mill with. I never refused to loan them any article, and they appreciated their neighbors so long as they were poor and in limited circumstances; but with his gang of boys, and their industry, it was not many years before they became well-to-do farmers, and get into good circumstances financially. They were a kind of self-supporting family. They would not pay out any money for repairs, as they were jacks of all trades and masters of none, and slashed into business with a determination to become wealthy. They even tried to do their own doctoring. The old lady was doctor for the whole family, and in many little ailments, such as breaking up colds, she would roll the children in a wet blanket, give them a little sage, or some other garden tea, and perhaps, in that way, saved many little doctor bills; so we can see there was some economy used in that family. The old lady had thus doctored her own family for many years, but on one occasion she failed. She had a son, a strip of a boy, who was lingering all one fall, and finally he got bed-fast. She still tried her skill with different home remedies to cure him, but he was getting worse, so one day I was called in to look at him, and they wanted my opinion of the illness. I told them the boy had typhoid fever and advised them to send for a doctor at once. They did so. Dr. Brandon was called and he pronounced it a dangerous case of typhoid fever. They lived in a small house, had a large family, slept two and three in a bed, and almost every one of them caught the fever and the house seemed like a hospital. The doctor finally said they could not get well all crammed up in a little house. Myself and wife were their main stay, as nurses, while they were in that condition. We went and sat up with them, night after night, taking care of half a dozen or more sick with the dangerous fever.

It was a cold winter and they got out of fuel. I sent my team along with some others to the timber to get wood. Their condition got worse and worse. I sent word to the Dunkard bishop that the family were in distress, and that as we had a family of small children ourselves, we could not spend all of our time with the sick ones. The bishop came and paid them some attention. They counseled with the doctor, and he advised that the family should be separated or they would die. They were placed with different neighbors and cared for. I took two of the boys to my house; they were the only two that were not down sick, although they were complaining, and would soon have been down if they had not been removed from the old house. A day was appointed by the neighbors to go to the house and do their butchering and have a general cleaning up of the residence. Some of the Dunkards met there, cleaned it up, and gave it a thorough whitewashing.

I did not belong to the church at that time. Old David Frantz came to me, laid his hand on my shoulder, and said "Chris., we know you folks have done a good deal for the sick ones," and added, with a smile on his face, "the horse that is willing to pull must always pull the heaviest load. We look to you to care for these folks' meat. You have got a cellar, and you can take their meat and salt it." I told them that I would do it, and in the evening I ordered my team, and took the meat and salted it.

It had a happy effect to move that family out. They soon all got well under the doctor's care and nursing, and in three or four weeks they were all well enough to move back on their farm, and into a whitewashed and garnished house, and the reunited family appeared to be very happy.

In the following spring while I was in the field adjoining Buckingham's farm, plowing, old Freddie came to me, and speaking of their sickness, and the good attention and kindness that myself and wife had displayed in their affliction, asked my charge for it. I said I made no charges, as I thought it was our duty to help our neighbors in times of sickness and distress. Well, he said he was very thankful to us, and that he wanted to be as good a neighbor as he knew how. I told him that was all right; that we might need help at some time, as we were all liable to be sick. It so happened that my wife was very sick for several months. She was not expected to live; the doctor had poor hopes of her. I had to stay with her night and day for six long weeks, and we were getting short of fuel. My hired hand did not know the way to our timber, so we sent a boy over to Buckingham's for one of them to go with their team with my man, and bring us a load of wood. The answer came to me, "We can't; we are going to haul brick for a foundation under our house." It seemed that he had forgotten the promise he made in the field, to be as "good a neighbor as he knew how." Afterwards, my wife induced me to go down to Buckingham and get his spring wagon to go to town in, as she was not well, and thought a spring wagon would be easier to ride in. They refused it, sending word that they did not want to loan it. I once went to borrow a hand press to make a little cider to drink. I asked the old lady for it, and she said they only got it for themselves, and did not like to loan it. I let loose on her, and said, "Mrs. Buckingham, you have forgotten that you was poor years ago. You even could not help yourselves, but borrowed pretty nearly all our farming tools and plows, and even our flour sacks to go to mill with, and now you have got independent, that you can get things for yourselves, but dont't want to loan them." I then walked off. Some folks can't stand financial prosperity. It makes fools of them. Buckingham and his whole family became tools for a set of hypocrites, became worshippers of liars, and lied to please their bishops. They came into court and swore to lies to please their leaders.

Frederick Buckingham was induced, either by the persuasion of these bishops, or by the devil's hired hand, or directly by the devil himself, to go into court and swear that he would not believe me under oath. I would, right here, ask all candid, good and moral citizens, whether such men could be trusted under oath. I would not, and could not, believe them. As old Bishop Wagner said, "He that has lied once will lie again, if he is pushed a little." But a number of them came to court and swore to what they knew was a lie, and it was done in a voluntary manner, without compulsion and with malice. We think there was a strong persuasion by the spirit of the devil, for we can read in the Divine Word of God, that He will send such evil men and women a delusion, that the may believe a lie and be damned. There seems to be a vast difference between lying and swearing to a lie. They are both sinful acts, but swearing to a lie is violating the laws of our country as well as the laws of God, and is a penitentiary act, if the laws of our States were enforced, but they seem to be very much neglected as well as the many church rules and laws.

Buckingham swore on the witness stand that my boys did all the work on the farm, also that the farm was grown up in weeds, which was a lie; all of the farmers knowing that my farm was cleaner of weeds than any farm in that section of country.

Now although those evil men and women have been trying to down me, they have received many favors from me as a neighbor; and because I have seen an abominable evil in the official board of the church, and contended for the right to prevail, it seemed that all the impudent, ignorant, devilish people, and devilish preachers hissed at me to aggravate me. It seemed as though the guns of hell were turned loose on a persecuted man, who was only fighting on the side of right, morally and religiously. It has not been a small affair for one man to stand and fight a legion of devils, and there are but few, perhaps, who can comprehend my sufferings. I suffered worse than many deaths, but many stood by and made light of this wonderful persecution. It was a feast for some of the hard-hearted lawyers, even some preachers made light of it, instead of coming to my aid. They appeared to want to see whether this Christian Girl could stand as much persecution as did our Saviour. One preacher even said, "Mr. Girl, you must stand it all if needs be; be nailed to the cross." I said: ' Mr. Chew, do you suppose you could stand all that, to be nailed up to the cross?" and he said yes, he knew he could if he had to, but he would not be nailed up without murmuring. I told him not to tell that around too much, or they would take him for an insane man. Lawyer Buckingham could have made a better case of that preacher's talk than what he made out of the Girl case. Well might we say with the poet:

"Be firm, be bold, be strong, be true,
  And dare to stand alone."

Please turn to Ephesians, chapter six, tenth and thirteenth verses:

"Finally, my brethren, be strong in the Lord, and in the power of his neighbor."

"Wherefore take unto you the whole armor of God, that ye may be able to withstand in the evil day, and having done all, to stand."

I shall try and expose all that had the impudence to come to court, and take the oath to tell the truth, and nothing but the truth. If the sheriff had put the obligation to all those who swore they would not believe me under oath, as follows: "You do solemnly swear to a lie in the case now pending in this court against Christian Girl, to lies, and nothing but lies, so may the devil help you," then they would not have perjured themselves.

THE SISTERS.

I will now have to expose some of the good sisters. We will call them good sisters, for we know that they have some very good qualities about them as neighbors, and in former years have been friendly and even neighborly to this unlucky family, but the poor mortal souls were overpowered by the stronger sex to give wrong counsel. It was proven at court that even some of those neighbor women had met at secret crossroads, and at each others' houses to hold their counsels. They were really influenced to take an active part in helping to confuse this once loving family, that they were once very social with, but those women became tools for the men that stood at the head of all this trouble, and they had timely warning not to listen to such devils as A. Bingaman, C. Funk and others, that had their hearts full of malice against me, but they were ready to be fed with the untruths, and some of them were influenced by those pious-looking devils to come to court and testify against me. I will say that none of those old pious bishops, or any of the devil's hired hands were ever instructed to go and speak peace to that confused wife and family, but on the contrary, they tried to keep them in a confused state of mind. I remember when I came home from Iowa, and found my wife in an unsettled state of mind, while I was explaining to the family what a time I had had with my crazy brother and his wife, who had been sent many untruths which made them obstinate towards me, my wife disappeared unnoticed. I asked my children where she was, and they would not tell. I spoke to my eldest son and he would not say where she went to. I told them to go and see where she had gone to, but they would not. So I started for Cripe's, the nearest neighbor. It was bed-time, and I called them up. Mrs. Cripe responded, and I asked if my wife was there, and she said yes, she had just gone to bed. I told her I wanted her to come and go home with me; that I would take care of her. She said, " No, Chris., you just go home." I told her that I did not want my wife to be under her influence, for she would not give her such counsel as she ought to have, and I didn't want her under her control; but she stuck her head out of the window and babbled with me and said, "you just go home and leave here." I told her she was not going to dictate to me nor to my wife, and that I would not go home until my wife went with me, but she stood and jawed with me until she got so mad that she fell into a fit. I said that I wanted to see my wife, and she must come home with me, or I would stay there. "You tell her to come down and go home with me, or permit me to come up and stay with her until morning, when I will take her home." She finally told me to come up, and when I got up she had a fit and got very sick. I asked my wife to get up and wait on her. My wife went and made her some tea of some kind, and worked with her about an hour and a half, rubbing her down to steady her excited nerves, and finally we go the angry woman quieted down and she dropped asleep.

That is way with some people. If they cannot have their own way, right or wrong, they go off into conniption fits. She was accustomed to having things her own way, as she had a very easy-going husband. He was sickly and she did about all the bossing there was done around the premises.

She had rather a bitter feeling against me ever since, and she came to court and testified to things she had not ought to, but she did it with a revengeful feeling, even winking to one of her own sex in order to attract her attention to notice how smart she could answer the lawyer's questions. The lawyer asked her "what more did Mr. Girl do to aggravate you after he came to your house to look after his wife?" She said, "he would pass me sometimes and call me smarty."

Good sister, I hope she will get her full share of admonition in this history, as it is not intended to destroy or tear down, but to build up. Come, brethren and sisters, let us look upwards. Heaven is above us, and hell below us; but I will show no partiality.

Mrs. Frederick Buckingham seems to be an industrious woman. She is a little like myself in some respects — she likes to gad about over the neighborhood and carry news. She likes to go somewhere where she can get a good dinner, tell news, complain about some of her neighbors and talk about the church members, brag about her chickens and pea-fowls, and her boy and girls, and boast of her financial prosperity. She also became a very active worker in trying to send me to the insane asylum. She made a remark like this: "What in the world will the people do now they have cleared Mr. Girl. A sane man, and he has come home to his family." She was maintaining that I was not safe to run at large. The poor old soul is not able to judge or govern her own family, for she had to let them out one winter for other people to take care of. I suppose she never saw an insane women only when she would look in the mirror. Ignorance, in its worst form, is worse than insanity, in my way of thinking.

When my wife and I separated I went to church with the hope of getting a chance to talk with her. After the meeting, Mrs. Wagner said, "There is Chris., if I was you I would go right by him, and I would not talk with him." She did as she was bid by this good old deacon's wife. How is that for counsel for a deacon's wife to give? I wonder if she can read the Word of God, or hear it read, where it says, "Blessed are the peace-makers, for they shall be called the children of God." She did all she could to separate man and wife, children and parents. The devil gets all such people unless they repent from their evil ways. "What God hath joined together, let no man put asunder." Mrs. Wagner had better read I Corinthians: "The wife is bound by the law as long as her husband liveth; but if her husband be dead, she is at liberty to be married to whom she will; only in the Lord." I make due allowance for ignorance. The law knows no ignorance, and we can read the laws of God so plainly that "a wayfaring man, though a fool, may not err therein." We will now drop the curtain on that sister.

I have something to say about Mrs. Daniel Wagner, the old-order deacon's wife. She had a hand in giving counsel to my wife and children, and helped to separate man and wife. On one occasion I was off to Ohio with my youngest son, and when I came back wanted to talk and reason with my wife. I went on Sunday to the old-order meeting, and I learned, by experience, that when you get into trouble with ignorant people, who are also full of self-righteousness and self-praise, you might just as well try to reason with a Japanese god, that is carved out of wood or made out of clay. As the Rev. Sam Jones said in his lecture at Decatur, Ill., "Worship man! what is man? made out of dirt." Well, it is said in the Word of God that man is made out of the dust of the earth. It seems to me that some of us are made of very inferior dirt, or muck. Well, it makes no difference what kind of dirt we are made of, we are assured that God has placed man and woman on this earth for a noble purpose, and, my dear fellow men, the question comes home to each and every one of us, are we filling our stations in life as one should, in order to be a light in the world. Suppose we, with all our Christian liberties and good literature and Christian training, should live out our Christian professions, would not that be better than to have the pulpits lined with skeptics as teachers. I say away with the whole of them to where they belong, to the devil's kingdom.

If you cannot get the skeptical preacher out by fair means, then use muscular power and kick them out of the pulpits, run them out of the country, or build a house for them where they may be kept until they repent of their sin of mocking their Creator, and destroying all the good that still exists in some creeds as they are called. Away with sectarianism. It is the devil's work. We will now give a word of admonition, through the Word of Truth, to this pious sister, hoping she may be benefitted by the instruction of our Saviour. A hint to the wise is sufficient. We would suggest I Peter, chap. iii. We hope the sister will ponder on this chapter. It will do her more good than running over the neighborhood speaking about things she does not understand: that is to say, " What will the people do now that a sensible jury has cleared Mr. Girl, a sane man, and pronounced him just a little too smart for his antagonist? How can we, as neighbors, have this christian man among us? Away with him. Crucify him." Another one will say, what evil has he done? I hope they will try and examine themselves, and see how they stand with their God.

We will name out another sister who was a close neighbor to me. It is Catherine Eshelman. She seems to be a sensible woman, but was carried away by bad influences by putting too much trust in man and not enough in God, and was made to take an active part in confusing my family, perhaps not willfully, but ignorantly. She put a wonderful trust in what old Johnny Metzker would say and do, and the boy preacher and C. Funk were her counselors, and it seems she was ever ready to obey their dictates and commands in the conspiracy.

She was in their clique. They used her as a telephone to convey their plans from my family to the evil party, who were planning how to defeat a christian man, and send him off to Jacksonville Asylum. Is that a proper place to send christians? Well, this pious-looking woman was almost as much confused as my family. She came to our house in former years to borrow such things as were needed in the household, but in late years she came to borrow trouble, and she got a good supply of it — enough to last her a lifetime. Her evidence was not much better than the boy preacher's. She was under his influence. They were all trained to speak the same thing. She was willing to go along with the devil's hired hand to Decatur, before Lawyer Buckingham, a very shrewd lawyer in the way of obtaining money. They laid the devilment before the lawyer by false representations against me. He told them that Mr. Girl was the one who ought to bring the complaint. They came home, and God, and themselves, perhaps, only knew anything about their plans of devilment. After they got home Mrs. Eshelman was sent to Mrs. Girl and made her believe that Buckingham said that she must go and enter complaint against her husband.

She said, no, she would not go, but Mrs. Eshelman coaxed a while, and finally made her believe that she must go, and they took her in the next day, and she was asked some improper questions, and she got mad and walked away from those devils; but they made up enough to make out a case of insanity, as they thought, before Judge Greer. I will now give some of Mrs. Eshelman's testimony on the witness stand:

Lawyer—"How long have you been acquainted with Mr. Girl, and what kind of a neighbor did he used to be?"

Answer—"He was a good, kind neighbor."

Lawyer—"Ah! He was a good, kind neighbor until your folks began to tamper with his family, and took secret counsel against him, and wanted to send him off to Jacksonville; then he got saucy?"

Ans.—"Yes."

Ques.—"Well, Mrs. Eshelman, did you think Mr. Girl was insane?"

Ans.—"I could not just say he was crazy, but he was very much bothered in his mind."

Ques.—"Did he come to your house often?"

Ans.—"Yes."

Ques.—"What did he want at your house?"

Ans. — " Oh, he wanted me to go and talk peace to his family. He said that there were a set of devils confusing his family."

Ques. — " Did you go to his house ?"

Ans. — " Yes, I did, but he forbade me to come there, and said he would throw water on me if I did not stay away."

I found out that she was giving bad counsel to my family, which was the reason I objected to her coming.

Mrs. Eshelman is a woman that talks very much, and Mrs. Girl said that Mrs. E. told her some things which Mrs. E. denied, and they both got very angry at each other. I was glad at that for they were too intimate with each other. Mrs. Girl got so that she thought more of her than she did of her husband, because Mrs. Eshelman would convey the devils' messages to Mrs. Girl and the children. She became the devils' telephone. Poor old soul. Her brother, George, told me that he had many arguments with her, and told her that she would get herself into trouble. He told her that Mr. Girl was not insane, and that he was competent to manage his farm, and he could make more money in dealing and hauling stock than all his boys, but yet she thought he ought to go to Jacksonville. He told me she would take no heed to his good counsel, but she took the counsel of those pious devils, such as the devils' hired hand, A. Bingaman, and the young outlaw, C. Funk. He that will not hear must feel. She was too lippy; she would not give me a chance to explain to her the mystery in the case. In order to be understood by her, I commenced writing to her, and she would send the letters back to me without answer to my questions. I will insert my letters on another page so that the readers can see for themselves what the object was in writing to her — to open her understanding, so that she might not be exposed with the hypocrites. Three of her inexperienced boys were induced to go to the court and testify that they would not believe me under oath, we think mainly through the influence of C. Funk.

We hope that this old lady has learned a lesson that she will never forget as long as she lives.

T. QUICKELL.

Here is a man I have had dealings with on several occasions, and in various ways. His name is Tobias Quickell, a brother to the George Quickell I have spoken of in a foregoing chapter, as having a church trial with him in York county, Pennsylvania. They are called very close dealers. There is a vast difference between a close dealer and a dishonest man. The Scripture says that we shall provide things honestly before God and our fellow-men,—we suppose it includes women as well as men. This Tobias Quickell is a pious-looking professor of Christianity, but is not strictly honest. I sold him a farm of eighty acres at his own offer, for a trifle over four hundred dollars. The tax was not paid at that time for the past year, but was about due. There was some little quibble between us as to who should pay the tax, but I finally agreed to pay it. After that an assessment was made for the following years, and in writing up the bond for a deed, Quickell, being a good scholar, formerly a school-teacher, managed to have that bond drawn up in such a shape that it made me liable to pay both the taxes, the one that was due and also the assessment to follow the next year, and he was willing to hold me to such unfair dealing, but I gave him to understand that he would get into trouble over such crooked ways. It was a very dishonest act, for the assessment was not in dispute at all at the time the sale was made, only the tax due at the time of sale. These are rather rough things for a pious professor to do, and I suppose because I had reproved him for his dishonest dealing, he became angry at me, and went right over into that clique of evil men, and was willing to testify that that he would not believe C. Girl under oath. This pious-looking professor took a very active part in the conspiracy against me.

I have had dealings with this man at various times; borrowed money many times; paid him as high as twelve per cent., and he always got his money at the end of the time named. What would a sane man expect from such a miser? He testified that he believed I was insane, and showed in many ways that his heart was full of malice. Thus one after another was induced to help to persecute me, and to confuse a once loving family. Can any sound-minded man support such hellish acts? The very devil is the author of this confusion and disturbance, and the editors are ready and willing to support such hellish deeds by publishing articles for the very author of lies, who comes in a sneaking way to establish his going by proclaiming to the public, as he says, a friendly offer. He says: "I am now past eighty years old, and for more than fifty years have been preaching the gospel of Jesus, and I am just as much interested in building churches, saving sinners and edifying saints as ever before." That sounds well for an old soldier of the cross, as he is termed, over eighty years old, who has made threats like this: "Sister Girl, you tell Chris. if he don't watch out he must go to Jacksonville. Tell him, but don't tell on me." Can the reader see any devilment in that pious-looking veteran? He says farther: "I felt like making a friendly offer. It is this: I still have a little money left, which I propose to give to poor churches to help them build. In giving I want it to do all the good possible. I desire it to serve a double purpose; first, to stimulate those who use tobacco to do well by quitting its use, and second, by helping to build meeting houses for those poor congregations who succeed in persuading those members among them who use tobacco to quit. They may apply to me, and I will cheerfully do what I can, with the help of the Lord, to assist them. This is not offered to hurt any one's feelings, but to urge a change for the better, and do that good which seems pleasing to the Lord. Pray for me, and may grace and love abound."

We wonder whether prayers would do him any good, with his load of sin. We think he belongs to the devil's kingdom, and he has a host of followers to that place.

Will the reader turn to Luke, chapter fourteen, verses fourteenth to eighteenth:

14. And thou shalt be blessed; for they cannot recompense thee: for thou shalt be recompensed at the resurrection of the just.

15. And when one of them sat at meat with him heard these things, he said unto him, Blessed is he that shall eat bread in the kingdom of God,

16. Then said he unto him, A certain man made a great supper, add bade many:

17. And sent his servant at supper time to say to them that were bidden, Come; for all things are now ready.

18. And they all with one consent began to make excuse. The first said unto him, I have bought a piece of ground, and I must needs go and see it: I pray thee have me excused.

Fear not, ye child of God. If the devil stands in the pulpit and throws his fiery darts, be not afraid, but stand firm for God, do right, and trust in God.

DANIEL WAGNER.

I will now speak of one of the old deacons, who is a pious-looking old hypocrite, who went into court and testified in favor of those leaders of the conservative order, after telling me of their crooked ways, and after withdrawing himself from them, and joining the old order, because they were corrupt in their rulings. The deacon and his son, Isaac, both spoke to me of selling their farm and going to Cedar county, Iowa, and joining with some brethren there who had withdrawn themselves from the main church. I argued the case with them, and I bid them to stand firm to their post, and take the Scripture for the man of their counsel, but they could not stay under such bad ruling any longer. I told them that there was a work to do to correct such bad ruling. There is a plan laid down in the blessed Word of God to handle such people, and if they would stand by me that we would soon overcome that evil element; and they said I could do nothing with that cousining ring. I asked them to do nothing against me, and I would show them what one man could do if he would put his trust in a higher power than a poor mortal man. They seemed willing, and even anxious, that I should proceed and fight their bitter enemies. They were willing to see the battle, but were too cowardly to strike a lick in the right direction. I called them drones and idlers. After they joined the old order, they were soon placed in the office of deacon. We will see what they did in keeping their words of not doing anything against me. The old man was influenced to go into court and give evidence that he would not believe me under oath. And now for his testimony:

Lawyer.—"Mr. Wagner, do you know Mr. Girl?"

Ans.—" Yes, I have known him for over twenty years."

Ques.—" How is he for truth and veracity?"

Ans.—" He is not good."

Ques.—" Would you believe Mr. Girl under oath?"

Ans.—" No, I would not."

Ques.—" Mr. Wagner, why is it that you would not believe Mr. Girl under oath?"

He was silent for quite a while, and the lawyer and jury were waiting for an answer. He sat there as though he was deaf and dumb. Finally, the lawyer said: " Mr. Wagner, you must have a reason why you would not believe Mr. Girl under oath?"

Ans.—" Well, I bought a cow of Mr. Girl years ago, and she did not prove to be what he recommended her. I paid him forty dollars for her."

Ques.—" Did you keep the cow?"

Ans.—" I kept her for three weeks."

Ques.—" Did you take the cow back to Mr. Girl, and did he pay you the forty dollars back again?"

Ans.—" O, yes."

Lawyer.—" That will do, Mr. Wagner."

I know that the cow was a No. 1 cow, and that she gave a large pail of milk when he took her away from the farm, and also when he brought her back. I made him believe I would take him up for perjury and slander. He got scared, and went off with his wife out west, I think to Wyoming Territory. They stayed there for some months. When they came back the court trials were not all over with, and the constable told me they had tried to dodge him, and that he caught some of them lying, some of the women even lying for their husbands in order to dodge the officer.

Now, this same pious-looking deacon, before he was induced by an evil spirit, told me, in his own house, that he saw that I was a terribly persecuted man, and that he passed many a sleepless hour in his bed thinking over my grievances. He pretended that he was one of my best friends, and he bid me God-speed in whipping out those that he hated so bad. He called the conservatives hypocrites, but gave in evidence enough to cause the brethren to become disgusted with such a deacon in their fraternity; but, perhaps, they are powerless, for he will deny these facts to them. Are those Christian principles, or the work of the devil? It will not take a very shrewd man to see through the devilment. I sent a spy to test him, and he spoke unconcerned of me, and spoke of what a time I had with those Dunkards up around Cerro Gordo, and he told the spy that it was a shame the way they had abused Mr. Girl. He said, "I have lived neighbor to Mr. Girl for over twenty years, and I never had a better neighbor than he was." What would you call such a man as that? Would you call him a snake in the grass? Yes, worse than that. A devil in the church, set up on a pinnacle as a pattern for the congregation. Great God! what will our Christian religion be in ten years more from now, if there is no change made for the better.

A stranger told me he had left his church, and I asked him on what account. He said, "They wanted to keep a man in the church who had stolen a hog worth forty dollars, and I told them that if they did not expel him I would leave the church. He was a man of means, and tried to be a hog thief and a christian at the same time."

I will now quote a passage of Scripture for the old deacon, and then will drop him, hoping that the Lord will pick him up and set him in the christian's path. He is serving the devil, and he will reap the wages of sin unless he changes his ways. May God help the unbelievers and the mockers to a better and a higher standard in the christian life.

Please turn to II Peter, chap. ii:

1. But there were false prophets also among the people, even as there shall be false teachers among you, who privily shall bring in damnable heresies, even denying the Lord that bought them, and bring upon themselves swift destruction.

2. And many shall follow their pernicious ways; by reason of whom the way of truth shall be evil spoken of.

4. For if God spared not the angels that sinned, but cast them down to hell, and delivered them into chains of darkness, to be reserved unto judgement.

8. (For that righteous man dwelling among them, in seeing and hearing, vexed his righteous soul from day to day with their unlawful deeds;)

9. The Lord knoweth how to deliver the godly out of temptations, and to reserve the unjust unto the day of judgment to be punished:

12. But these, as natural brute beasts, made to be taken and destroyed, speak evil of things that they understand not; and shall utterly perish in their own corruption;

13. And shall receive the reward of unrighteousness, as they that count it pleasure to riot in the day time. Spots they are and blemishes, sporting themselves with their own deceivings while they feast with you;

15. Which have forsaken the right way, and are gone astray, following the way of Balaam the son of Bosor, who loved the wages of unrighteousness;

16. But was rebuked for his iniquity: the dumb ass speaking with man's voice forbad the madness of the prophet.

I will now show to the reader some more of the opposition I had in fighting my enemies, who were trying to do me all the evil they possibly could, but in a sly way, so that the world should not suspect them, and they tried to induce everybody to go against me — tried every way that the devil dictated to them to injure my reputation; at the same time they would stand in the pulpit and pretend to their hearers that God is an allwise being, one that knows the very intents of the human heart, and God wants us to worship him in the spirit and the truth; that we cannot deceive God, and a liar cannot enter the kingdom of heaven. The wages of sin is death; and they were very particular in laying before their hearers the ordinances of God's law at the time of their communion meetings, before going to the Lord's supper as they termed it, and they would refer to such chapters as warned them to examine themselves.

ISAAC WAGNER.

I will now show to the reader that I have in all instances warned all who have become subjects of this book by letter. Some of them I have warned for many years. Here is a deacon, a son of "Old Older," who gave in such bright evidence to try and break my oath in the cow-business, and it seems as though his son Isaac, is willing to sustain his father in his false evidence. We hope the bishop of that church will not suffer such things to be smuggled up in their church without a trial of investigation. It will render both father and son unfit to belong to any christian organization unless they confess their sins before God and man. Isaac is a man who thinks himself competent to judge for himself in financial and religious matters. He was ready to condemn the conservative party, especially the cousining ring. He said there was no good government in that element of the church, and that he was going to leave them, and go with the old order. I told him he had better not do that, but stand by me and we would correct such bad ruling. He laughed at the thought of my working against such a force, and said they would overpower me. I said, no, they would not. I told him he did not believe the Word of God, where it says that the christian believer shall not be persecuted above what he can. Well, he turned "old older," and some more followed his example by rising to their feet, and organized themselves into a class. Isaac Wagner and his father were elected to office as deacons, and old Jacob Miller as their elder. Isaac Wagner became an active worker in trying to build up the "old order" church by quarreling with those same persons that I condemned as liars and hypocrites, and before he left them he knew many things and said many things that were hard against them, but he did it in a sly way. He was too cowardly to come out and fight them in the open field. He was willing to carry news, and that just made me much trouble. So one day I met him on the road, and told him he had to stop that kind of work; that he must stand on one side or the other; that he could not carry water on both shoulders. I was on horseback, and he was cutting weeds and had a hoe in his hands, and he sassed me, and I, as quick as thought, got off my horse and started for him. He leaped over the hedge like a deer, with his hoe in his hands, and I went on about my business. After that he and his father were willing to stand in to help those ornery ones to scream out their mischief. Can we call such as them true friends?

A butcher bought some cows of his father and his brother-in-law, and they were to deliver them for so much money, and in bringing them to town they tied one behind the wagon and it got stubborn, and they pounded and dragged it in such a way that they finally had to load it on the wagon, which rendered it unfit for beef, and they met and drew their money. The butcher, not being satisfied with such a delivery, proposed that they should lose half of the value, as it was delivered half dead. He sent word to this Isaac Wagner that he would bring suit against them if they did not come and settle the matter. The young deacon began to bristle up to the butcher and make threats. He said to the butcher, "if you sue my father, I will come and swear that you told us to bring them dead or alive." The butcher said he would give him a thousand dollars if he would swear to that. He could not scare the butcher with his unreasonable threats, and he went home after telling another lie on the top of the one he was going to swear to. He was asked whether his brother-in-law was in town, and he said he was not, but the butcher saw them together in town the same day.

Will the bishop correct such deacons. If he will not, then I would say to him he is no good shepherd. The world is talking about such ornery church members, and well they may. May God pity such weak members and take them out of the way of those that would do well, and might be good workers in the church, for the world to look at and patern after. O, may they repent and do better. They boast of heaven and happiness and have nothing but the empty husks, — only a form of worship. Cease to do evil and learn to do right, is our Saviour's instructions, and be saved.

JACOB WAGNER.

We now speak of one of the now-deceased joining elders, namely, Jacob Wagner. As it was my privilege to call on the joining elders to investigate this trouble, I visited him at various times, but he declined having anything to do with the matter. After hearing my grievances he acted rather sly and obstinate, and walked by me without speaking a word. I wrote very personal letters to him about his slackness of duty as a shepherd. I asked him and his wife to come and see me, but he refused to do so after I had been acquitted at court and pronounced sane. I met Mr. Wagner on the sidewalk in Decatur, and asked him "what do you think now?" He said, "Well, Chris., there is something wrong somewhere." I had told him what was wrong long before that, but that was all I could could get out of the old bishop at that time. Sometime after that I saw him talking with Deacon Funk, and I walked up to him and said, "Wagner, what did you mean by it when you said there is something very wrong somewhere?" He said that he meant that all the wrong was between me and my family. I told him that it was not true, and that he knew better; and I left him and wrote him another very personal letter. I afterwards went to his house. The old lady bid me be seated, and pretty soon the old bishop came in, and I told him what I had said to his wife. I told him that I had written some very personal letters to him, and that I was ready to make acknowledgments if he asked them of me. He said he had nothing to ask of me. It was pretty near night, and they asked me to sleep there. I unhitched my horse and stayed, and was treated kindly. I talked very free about the evil men and women who had falsely imprisoned me, and how treacherously they had treated me and my family, and there was no contradiction made by either the bishop or his wife. They seemed as though they were afraid to take any action in the case.

It was not very long after this that God called them away to eternity, the bishop dying first, and his wife followed him in a few weeks, and they are now in the hands of a just God. We have nothing evil to speak of them, only we think there was a lack of the Christian courage so much admired in the teachings of our Blessed Saviour by all true people. As the poet says in the hymn eight hundred and four:

"Be firm, be bold, he strong, be true,
    And dare to stand alone;
Strive for the right whate'er ye do,
    Tho' helpers there be none.

"Nay, bend not the swelling surge
    Of fashion's sneers and wrong,
'T will bear thee on to ruin's verge,
    With current wild and strong.

"Stand for the right though falsehood rail,
    And proud lips coldly sneer,
A poisoned arrow cannot wound
    A conscience pure and clear.

"Stand for the right, and with clean hands
    Exalt the truth, on high;
Thou'lt find warm sympathizing hearts
    Among the passers by."

LEONARD WAGNER.

I will now speak of one of my neighbors. Leonard Wagner is a deacon in the Okaw Church, and became one of her counselors. He is a brother-in-law to A. Bingaman, and pretended that he stood neutral, but I mistrusted that he was a tool for his brother-in-law, the devil's hired hand, so one day I made him show his colors by coming down on him. He stood in the open field against me, and was very impudent and saucy, and even abusive to me, but of course I could hoe my own row with him, for he was a fool and a tool for his ornery brother-in-law. I afterwards met Wagner on business, and tried to buy some corn from him. He was short in his talk, and would not sell me any corn. I told him he was short and snarly, asked him what was wrong, and why he did not reply. I told him that I expected to move on to my farm in the spring, and that I wanted to be friendly, so that I would have some neighbors, but he said he would not neighbor with me. I then asked him if I had ever done him any wrong, and he said I had been very saucy to him. I told him it was only tit for tat. That he had intimated that I was insane, and that I had just as much right to say the same of him. I asked him to reason with me, and told him that an intelligent jury had pronounced me sane. He answered and said, "You don't act like a sane man, going round calling people devils and hypocrites." I told him that I wanted to call people by their right names. His wife and mother-in-law heard the most of the conversation. I said to him, that I wanted to be a christian; that I thought I knew my christian duty. I then asked him if he could forgive me of all the wrong that I ever did to him. He would not answer, and I asked him the second time, but got no answer. I then told him that I was cleared before God and man and it did not matter to me whether he forgave me or not. I told him I could do no more with him, and "if you pray the Lord's prayer, you had better stop and consider what you are praying when you come to where it says 'forgive our debts as we forgive our debtors.'" We assured him that the malice that he held in his heart must be rooted out before God would accept him in His bright and glorious heaven above, where all is peace and glory among the angels. You cannot enter with stained garments on; they must be washed white as the driven snow in the blood of the Lamb, then God, in His infinite mercy, will assign you a place among His glorious angel band.

M. M. ESHELMAN.

This man is a natural-born fool. He wrote an article for a paper intended for my benefit, and I will now give it in my book: "A spiritual wind-sucker may be a necessity in the universe, I will not say he is not. He may be an instrument to increase patience and charity in those who are born of the spirit, but his methods and mission are not to be envied. And now, why dost thou afflict thyself, dear brother, because you cannot run in the ministry before you have learned to walk. Just why you should be cast down in your mind, or why you should feel ill towards those who long practiced before they could run, is not quite true to those crumbs. You had better get down to solid work, read, study, pray and meditate, and the sinews of your heart and soul will grow strong, so that you, too, may run a swift race. These crumbs have no use for words and letters which say ugly things about brethern and sisters. You may save your postage if you are tempted to inflict on crumbs your doleful tale, your envious soul, and your inflamed mind, because, forsooth, that at some time and in some place your brother was in your way. These crumbs have no need of evil surmisings. The story-peddler is a popular person in some places. It seems there are always some individuals standing on every cross-road, figuratively speaking, with ears open and grinning countenance to hear the latest gossip in the land. If the story-peddler has a good word for the listener's opponent he is not heard with pleasure."

Now, it is plain to see, that this was written for me. He charges my letters of being full of ugly things and vile stuff. If they are of such stuff I would like to know what kind of religion you are possessed of? You have put me down as a vagabond of the earth, and you want to drown and kill me from the face of the earth. Now, if you had any religion of the Blessed Saviour, you would have to meet your fellow-man that is in trouble, and administer to his grief and stricken heart; but no, you possess the spirit of the devil, and it teaches you to write those insinuating articles and have them printed, and send them broadcast over the land to try and crush me, and prejudice people against me. If the contents of your book, called the "Two Sticks," is full of such stuff, it will not be read with pleasure among enlightened people. You had better take it among the heathens, for they will not know any better. That may be the idea that has struck you, and is the reason you have come here among these religious fools to sell your "Two Sticks," Some of you people are just like a lot of dogs—wherever the big one goes, there all the little ones follow. All the little, ornery, long-bearded, flat-headed, good-for-nothing curs are to be found around the rotten cesspool of Cerro Gordo Church. Where the carcass lays there the stench is great.

Mr. Eshelman, if you had the true religion in your heart you would have answered my letters about exchanging books. Why did you not want to exchange books? It was that devil in your heart that told you not to do it. So you thought I was insane and could not write a book, but I will show you that I can write a book, and I can do more. I will expose you and all of your corrupt devils besides. You Dunkards are selfish; you have preached and prayed to my family until you have distressed them, and almost drove them insane, and then, instead of giving them a helping hand, you wanted to go into Mr. Berry's house and pray some more of your vile stuff to that poor, distracted daughter of mine. It moves my heart with emotion to think that you and that long-bearded fool of a Cripe had no more sense than to think that you could do a poor, sick girl any good with your prayers, when there is nothing more abominable in the sight of God than prayers from the lips of such men as you. Now take heed that you do no harm when you are walking up and down in the hearts of those people as you made your vain threats of bigotry, for that is all it is. It seems as pollution, you were so set in your blind ways that you overlooked good judgment. You might have known that my days were but few at best, but you seemed to delight in your persecution; but I have strong hopes in my God that he will reward me for all of my trials, and finally save me at last, and that is more than he will do for you hypocrites, for you will never see heaven if you live to be as old as Methuselah, for the longer you live the more sin you commit, and the more your deeds will damn you to hell, if you do not repent of your wicked ways.

JACOB DEARDORFF.

I will now give a brief statement of the trouble I had with my former tenants, Jacob Deardorff and son. The son is a spoiled boy. His mother says that the time he was unfortunate and lost his limb on the railroad as a brakeman, he was nearly dead, and they have petted and spoiled him. It is well known that when these people moved on my farm they advocated my cause very strong and condemned my persecutors. I made my home with them until they asked me for the farm for the fourth year. All was right with our settlements, which we had every year, and we always settled without a word of dispute. It could be, and was, proven that they had said to the citizens of Macon and Piatt counties that they had never had a better man to deal with than C. Girl; and thus we got along nicely until they got refused for the farm for the fourth year, and then that wooden-legged, unruly boy commenced to blow, and the old man commenced to blow, and they got the old lady riled up against me, and the daughter began to make sour faces at me, and I was no more welcome at their house, just because they could not have the farm for another year. I tried to reason the case with them, — that I wanted the farm myself to get my boys to farm in order to make more money to pay off the incumbrance upon it, so as to save it from being sold under a mortgage claim. I also told them that I wanted to get my family reconciled, but they would not accept my good reasoning, but kept on blowing and abusing me in every possible way, intimating that I was a thief. They crowded me out of the house, out of my reserved stable room, and out of my wagon shed. The old man said, "take your d — d old carriage out in the road." I did so, and afterwards made a shed for it on my reserved pasture lands. He demanded a settlement, and we appointed a time to settle before

Tommy Albert, in Decatur, as he wrote and held the articles of agreement. I agreed to meet him at Albert's house at one o'clock. I went there at one o'clock, and Mrs. Albert told me her husband would not be home until evening. Deardorff got angry, and went to Long Creek Station and sued me before 'Squire Rucker for a settlement, and as I was not ready to appear for trial on account of an absent witness, the 'Squire granted a continuance of one week, and I proposed to settle the matter right then and there by arbitration, and my offer was accepted. He chose a man by the name of Harry Wright. I chose Mr. Cochran, and they chose the third man, Mr. Nowland. Mr. Cochran wanted to know whether we would agree to stand by the settlement they made for us. I spoke up at once, and said, certainly, and Deardorff said yes. They then went to work and got our accounts. Deardorff's account was, I think, sixty dollars and thirty-five cents. My account was upwards of seventy dollars. The arbitrators gave him six dollars and thirty-five cents, and we were to pay the costs, half and half, which was three dollars and twenty cents. I pulled my money out and laid it on the 'Squire's table, but Deardorff would not take it, but went to town and saw Lawyer Buckingham, who induced him to take the case to court. Deardorff did so, and he made his bill against me for one hundred and ninety-nine dollars. When the case came before the court, lawyers Bunn and Park made a plea before Judge Hughes, and stated that the matter had been settled by arbitration before 'Squire Rucker, and the judge dismissed the suit.

I think the shrewd, but not wise, Buckingham, who belongs to the Baptist Church, must be a pillar and a pattern for a church to look at. If the pastor of his church can make a christian out of such lawyers, he is a good worker in the Lord's vineyard. It may be that the pastor is not after his soul, but it is possible that he may be after that which damns the soul, as we read that the love of money is the root of all evil. Will the reader think for one moment of the devilment of Deardorff's unjust bill of sixty dollars and thirty-five cents against a true and honest account of something like seventy dollars, and then go back on a committee's work, and raise their unjust account without any further dealing to the sum of one hundred and ninety-nine dollars. There would have been just as much justice on their side if they had raised it to one thousand dollars. Mr. Deardorff declared he would make it hot for me. If he believes that that there is a hell, we think he is carrying his own brimstone with him to his final abode to last him through eternity, unless he will repent of his unfair dealing. Lawyer Buckingham supported him in such a hellish work. Not satisfied with the judge's decision they went to 'Squire Curtis, and brought suit the third time. The old cripple was willing to hear the case, I suppose in order to make a few dollars, after the arbitrators had given in their evidence in favor of me, and had left it. The Deardorffs were put on the stand, and they testified that they had told the arbitrators that they would not stand to the settlement unless I brought in a fair account, and they swore to the most unreasonable things that the foreman, and all of them, and none of the outsiders, ever heard of before the men made their decision. The old 'Squire gave them a judgment for seventy-four dollars, and costs. I put the case back into court. The 'Squire needs some recommendations. I will let Lawyer Mills recommend him, as he became my attorney to handle the case. In talking to Mills about the unfair suit before 'Squire Curtis, he asked me who their attorney was. I told him Buckingham, and that Bunn was my lawyer, and he said that there was no lawyer who could win a case before Curtis against Buckingham. This is a good recommendation for a justice of the peace. I wonder what church he belongs to? Mr. Mills neglected the suit at court, so he is no better than the crippled 'Squire. He will make another good pillar for some church. Deardorff got a judgment against me which amounted to the sum of $119.95. He was in a hurry to have Constable Dillahunt serve it and levy on the last unincumbered property that I had left on my farm, namely, horses and plows, carriage and corn. Dillahunt made his brags that he had taken about a thousand dollars worth of property to make the amount. Deardorff went to the farm where the sheriff's sale was to be, I suppose, to buy some cheap horses, but the sale was stayed by me paying off the execution a few days before the appointed time, so Deardorff was defeated in buying cheap property. George Stair had a note against Deardorff of many years' standing. Stair said, if he thought Deardorff would not get mad at him he would let me have the note, and shove it against the judgment. Who would pity old George Stair if he would lose that note. I was feeding two carloads of cattle. I tried to get Deardorff to cut up my share of corn early in the fall, but he would not do it for spite. I asked him if I could cut up my share of corn to feed my cattle, but he objected to that. I then offered him ten dollars an acre for his share, but he would not take it, but finally offered his share of forty acres at fifteen dollars per acre. He told me that "Those Dunkards were right when they said you was crazy." Well, I told him, that if I paid him fifteen dollars an acre for his corn the people would have reason to think me crazy. They lost several hundred dollars by not taking me up at my offer. I went to Mr. Phillips and bought corn for ten dollars per acre, that had a third more corn than his. I told them they had better go and join the cousining ring of Dunkards, and get the boy preacher to baptize them. They even tried to prevent me from getting my own feed in the granary, declaring they would throw it out if I did not take it out.

One evening I went to Mr. Phillips on some business. The old man Deardorff and son were on the wagon at the barn, and they made faces at me, as I rode by on my horse, but not a word was said.

They were the only two persons I saw on my way that evening. On my return, along the east side of my farm, where there was a hedge fence with plenty of hedge apples upon it, all at once the apples came flying thick at me. My horse jumped; I checked him and turned around and said, "Who are you?" but there was no answer. I leave the reader to judge who the cowards were, and will call all such prairie wolves, as I called C. Funk and others. When the Deardorffs began to get bitter at me I begged them for mercy, and to listen to reason. I said, in a mild way of reasoning, you will hurt yourselves and your reputation, for I will prove myself the innocent party in this conflict, and begged of them to not go hand in hand with these evil men and women; but my pleadings for mercy were sneered at by them, and they laughed at me. Even Mrs. Deardorff sent a little boy out to the granary to see what I was doing while I was getting some of my oats to feed my horses under a shade tree, after they had crowded me out of my stable room. The little boy came to the granary as he was directed to do by the old lady. I talked friendly to the little chap. After I had several bushels sacked and carried out, the old lady shouted sharply at the boy to come into the house. When he got there she asked him what the old devil was doing. I don't know what the answer was that the boy made to her. I think it was on the same day that the old woman was boiling soap on my reserved land, and I happened to pass her on my way to pump some water for my stock. I said to her, "Mrs. Deardorff, I am no devil, but a terribly persecuted man by a set of lying devils." She made no answer, and it seems that silence gives consent. She was a little sorry that she was so rash in her expression to call a persecuted man a devil,— the one that bound his broken limbs and bathed them with liniments, who was bruised up by a cruel outlaw of a lying son-in-law. At that time the old woman took great pains to condem these brutal people. Old Jacob sometimes almost grit his teeth, he got so mad at the way my enemies treated me.

One Sunday morning two of Eshelman's boys came to his house, and I tried to reason with them to show them that they were under the influence of C. Funk, a regular outlaw. The big, overgrown impudent had the impudence to call me a liar. The old man could not stand that, and he drove the gang off from his house, but he afterwards called me a liar, a devil, and a dishonest man, and even a crazy one. Well, I will let the reader judge for himself who the crazy one is.

I had loaned this man Deardorff one hundred dollars before I started for Florida, and when the note became due, and past due, I became hard up for money; and asked him for its payment, and he got mad. He did not pay the interest, and so I bought the interest on my account when he sued me for a settlement. I have had many settlements with citizens both in Piatt and Macon counties for nearly twenty-two years, and never in my life was I sued for a settlement before. He sued me for spite and self-gain. I am thinking the unjust treatment will be a loss to him in the end. It seemed that he and his outlaw of a boy took great pains in trying to bemean me and break my character down. I understand that he went to an old, good, honest citizen, and he began to blow against me, to try and belittle me. The man listened to him awhile, and then asked him if he thought he could ruin a man's reputation by talking,—a man who had lived in their neighborhood for so many years, and had showed himself a good man by his home conduct. I gave Deardorff to understand that he would hurt himself to attempt such a thing.

This was my only means of fighting so many enemies. I might, perhaps, have whipped some of them by muscular power, but as that is not my favorite way of fighting — it is carnal nature's way, a kind of a doggish way of settling difficulties — I will try and take the Saviour's way as much as it lies in my power, hoping that when these remarks shall reach the guilty parties, they may repent from their evil ways, and learn a lesson not to be tools and act as fools for others, and come over to the innocent party, and not disgrace themselves. They are hated by God and man. We read that the christian will be hated by man — that don't matter so much, but if God hates them all earthly and heavenly hopes are gone. Let us act as wise men and not as fools.

When this family came on my farm they were fond of boasting of their city manners, and still they were very jokey, and they wanted to pick me out a wife, so they brought out from town their widowed sister, and planned to throw us together in society, and when I came home they could not wait until I put my team away. They stuck their heads out of the door and windows, and shouted, "Come in here, we have got somebody here; come in and see." I walked in, and one said, "See here;" another said, "Look there, see what we have got for you." Great God, what refined manners!

Everything went along very smoothly between us, and we all enjoyed ourselves for some time, until I refused to let them have the farm for the fourth year, and then it seemed as though the evil spirit took possession of them, and they did all they could to aggravate me and beat me financially. But God is good, and will deliver all his children from the hands of the wicked in the end. My prayer is that they may ask God to forgive them, and may they live better lives in years to come.

WILLIAM HEIL.

I will now give a few pages on account of my late tenant's conduct and actions towards me. He was introduced to me by A. J. Rainey, as a good farmer, and he also spoke very highly of Mrs. Heil — that she was as neat as a pin. I went out to the farm the same day they moved, and helped them to unload their goods, get some wood ready for a fire when their stove was up, etc. We worked till bedtime in the evening, and then went to Mrs. Marker's and stayed all night with them. Mrs. Heil did not come with them, but had gone to Decatur. When she came back I asked her whether she would board me awhile, as I wanted to fix the garden fences, and make a general improvement about the house. She was anxious that I should make the improvements, but refused to board me. I told her I would pay her for my board, but she refused to board me. She wanted me to fix up things, but I told her I could not do that without a place to stay, and they looked at me as if I was a prairie wolf on account of Deardorff's slanderous talk about me. I went to Mr. Heil and asked him if he would rent me the summer house, and he seemed to be afraid to let me have it. I made a proposition to him, that I would rent the house for one month at a time, and at the end of the month, if it was disagreeable, I would move away. After the month was up I said to him, "Well, Mr. Heil, how is it now; can we stay any longer?" He smiled and said, "Well, if you don't get any more insane than you have been thus far, you can stay," but before he permitted me to stay he said he would see

Rainey. I told him I was sure that Rainey had no right to object to it. Heil got it into his Dutch head that Mr. Rainey would have to be consulted about it. I told him that Rainey would laugh at him. I went to town and he wrote a few lines and sent them in by me to hand to Rainey. I did so, and he read the note and laughed, and said that he was not Heil's guardian, that he could rent me everything for what he cared. We got along together very well until after harvest, when he asked me for the farm for another year. I told him that I would farm it myself; that I wanted to get my family together and get them reconciled, and that I could not keep my indebtedness paid up, and that I must make more money in order to pay the illegal debts that had been forced upon me. He asked the second and third times.

About that time C. Funk and others found that he was getting crusty at me, and they were not slow to urge the jealous man with their lies as they did the Deardorffs. C. Funk made up a lie to get them angry at me, by saying that I said I did not get my share of the grain that was raised on the farm, and they were ready to believe him. So with Heil. He was ready to believe any lie against me because he had his heart full of malice. Reason had no influence on his rattled brain, but he became wild and irritable and even ugly, and began to make villainous threats that somebody would get their heart stamped out of them. One day he was in the stable and I and my son Charley were in the feed room, and he talked very saucy and impudent. I tried to reason with him, and Charley Girl shall be my witness if he tries to deny this. I told him I was preparing a history, and that certain ones would become subjects of it. In an important way he said, you had better give me a few pages in your history. I told him I would do it at his own request. He said he wanted me to write the truth, and I said, "O, yes, I am noted for that," I am penning it down this moment to be put in print to stand forever open before the public. Reader, this is the only way left me to break down the evil influence that was brought against me by so-called Christian professors. "He that giveth himself in evil hands shall perish therein." Heil is young yet, and has, perhaps, many years to live. It is to be hoped that he may learn a good lesson not to be persuaded by a set of outlaws.

Hoping my own young and inexperienced boys and girls, and others, may also learn a lesson, right here we will show some more of Heil's ugliness. He now became very jealous against my son Charley because he came home and commenced farming. They both met at the stable and he said, "Charley, you are a darned fool to come home and farm your father's poor land." Charley had been imposed on by him before that time. He is a good-natured boy. He gave Heil an answer to his impudent and insulting talk. The boy said to him: "Well, Heil, there are a good many who would like to get a chance to farm this poor land if they could only get it." After he was refused the land the third time, he delared "that he did not want my poor and stony land," although he had raised a better crop than he had ever raised before in his life. Poor as the season was, he raised over eleven hundred bushels of oats, and about two thousand bushels of corn.

He kept abusing Charley until the boy could bear it no longer; making threats to others that someone would get their liver stamped out. We naturally supposed that it referred to us, but as the saying is, "a barking dog never bites," we did not take much notice of his threats, but attended to our own business. Charley was plowing some wheat ground, and as we were scare of feed, I asked him whether he had any objection to us taking several rows of corn alongside of where we were plowing for wheat. I told him that we wanted some corn to feed a pig or two, and to use the fodder for the horses. Told him it was nubbin corn, and it would save him gathering that much and hauling it off to market, but he would not consent to it. I went and cut off a small load and took it to the stable. and while I was husking some of the nubbins of corn for my pigs Charley was carrying some of the fodder in to give to the horses. Heil came up to the wagon and commenced to make threats of prosecution for taking my own corn. He came close to me, made up his fists at me, and threatened to whip me.

I tried to reason the case with him, but he kept on calling me a liar, and declared that I was crazy, and that he would whip me. I told him that as far as that was concerned he need have no sympathy, for I had been proven a sane man by an honorable jury. I told him to keep his hands off me. Charley could not stand it to see his father abused in that manner. He stood about one rod from us, and heard Heil abuse me, and spoke to him, and said, " Who are you calling crazy ? If you think my father insane you had better not pile on to him." He then said to Charley, " Maybe you want to take it up for you father," and struck at him. Charley was too much for him. He began to back him up, and Charley made every stroke count, for I could see sore bumps on Heil's head, and his face began to get bloody. He was knocked flat on his back, and was badly hurt.

The old bulldog, who is so old that he has scarcely any teeth, and is nearly blind, got excited and commenced barking, and Heil, feeling the want of help in the struggle, tried to get him to join in. I and the bulldog were the umpires of the battle. The boy said to some of his associates that he enjoyed the fight. Heil was taken to the house by his wife, with a bruised head. It was noon, and when we were in the house eating he made threats that he would take the axe to the son-of-a-guns, but made no attempt to do so. He would have got a warm reception, for the boy was ready for him at his own game.

We then left the farm but our goods were still

there in the house. I wanted to secure my goods by putting a new lock on the door, and I was putting it on when he came and ordered me to stop. I stopped and went off, and the goods were exposed to thieves.

Early in the fall I walked into the orchard to get an apple to eat. Heil came out and said, "Come in, I am ready to settle with you." I walked in. I had my book with me, and he got his book. There was only a few dollars between us. I let him have his own way. He did all the settling and I submitted to it, and put it down and the date of settlement. Then I was ready to leave the house and go about my business, but he told me to hold on for a while. He was writing out a statement about our settlement. He did not read it, but got up from his chair and said, "Now, I want you to sign this paper." I asked him what for, and he said about our settlement. I said all we needed was to put down, the date of it and the amount due, and I have that down in my book. "He says, "well, I will make you sign it," and he began to make two fists at me, and tried to force me to sign his paper. He said, with as savage a look as his old bulldog could make, "Come right up here and sign this paper." He had made charges for some things he said he would not charge for, one of which was for cleaning out the cistern. He had told me if I would get some cement he would help me to fix it without charge, as they would have the benefit of it. He then said, "you will not go out of this house until you sign this paper." I began to look for a chair with which to defend myself. I said, "you will make me sign that paper against my will? No, sir. This will be a bloody kitchen before you make me sign that paper." He then began to lose grit. I told him that he could drive an ox to water but could not make him drink.

The reader may wonder how I got away. I stood my ground until he cooled down, and then I walked out and the coward followed me and abused me. I walked through the gate and left the gate open. His wife came out after him. As soon as I was outside the gate, off came my coat and it was tossed on the fence, and I said to him, "Now, come right out here, you infernal coward, and I will whip you in a minute." Then he backed out.

It was his request that I should give him a few pages in my history. I hope he will benefit by it, and never again mistreat an old man as he did me, and have a little respect for old age. If those things do not come home to him in his young days, they will in his old. It was nothing but spite work from the time they were refused the farm for another year.

If I was a tenant, and asked a landlord for a farm, and he gave me as good a reason for not wanting to rent it as I gave to him, I would not stand and quarrel about it with him, and especially with a crazy landord, as he said I was.

May they resolve never to abuse another landlord as they did me, and I hope they will repent of their evil deeds.

S. FOUTZ, J. WISE, and J. ULERY.

Leonard Foutz.

I will now speak of Leonard Foutz. He is a deacon in the Okaw Church. He seems to be an industrious old deacon, and some years ago he and his wife had a great deal of trouble about the evil ruling in Metzker's Church, and even had some very serious church trials before that precinct, and he and his wife seemed to entertain a hard feeling against the majority of that cousining ring, and they seemed to stand by me many years in putting down unjust proceedings in that ring, but, alas, they turned their backs when the battle began to get hot; they flinched, and I was no more welcome at their house, but they turned bitter against me, and even declared that I tried to destroy the Dunkard's Church, and I think they became active members against me, and they went hand in hand with my persecutors, blindly led along by a set of liars. We think such idolatry and combination should be broke up, then the christian religion would shine forth in its true light. May all religious people give the warning sound, if not in a history, in the way of lifting their voices against the great evil that is now praticed among christian professors. I hope they will remember that God is not to be mocked.

Johnny Wise.

I met Johnny Wise at the annual conference, and shook hands with him, and I opened the subject of my persecution. He seemed very friendly and sympathetic towards me, and I approached him like this: "Johnny, why don't you answer the many letters I have written to you for years?" He said, "Brother Girl, I have answered all the letters you have written to me." I told him that was strange, as I had received none of them. Now, I will not intimate that he did not tell me the facts in the case, but we will see how these things pan out. We will see whether I was earnest in my pleadings, for soon afterwards I lost my suits at court, or before Johnny Metzker's court, where he sat as judge himself, the cousining ring were the jury, and the devil's hired hands the lawyers.

Of course, a christian man had no chance left but to write a history and lay it before the world in its real nature, and Mr. Wise ought to have been wise enough to see the point where that cousining ring was drifting to. I said, "Johnny, what would you think of brethren who would go beyond the gospel of the 18th chapter of Matthew, where it says, "take one or two with you, so every word may be established." I told him that happened in our church, even with me. Why, the Scriptures are plain on that. I told him that some were called by me to go and see a member, and they went out of their way to bring a young preacher along; and he said, "I hope you have not got any brethren who would violate the rule of the 18th chapter of Matthew — such a plain gospel." I told him that there were two of the old bishops who brought the tonguey young prophet along some times. I call him the devil's hired hand. I told Mr. Wise that I sent Bingaman home; that he made himself entirely too useful. Johnny Wise may be a good man, we hope he is, but we really don't know whether he is or not. What I have written is not to try and hurt him, but only to wake him and others up to our christian duties. Let us all take warning, and take heed to the laws of God. Let men say what they will. A hint to the wise is sufficient. Judge ye what I say.

Jacob Ulery

is an old bishop belonging to the Okaw Church. He always seemed to be sympathizing with me and my family. I don't think he felt like taking a stand against the persecuted man. He seemed to want to stand in this fight between right and wrong. I called on him to go with me to the boy preacher to try and set him right according to the 18th chapter of Matthew, but he refused to go with me. He made the excuse that he was lame, that he had a sore foot. He was out hauling shock corn at the time he made the excuse. I expect he thought a poor excuse was better that none. But we are not excused from our christian duties, according to my understanding, as we can read, "he that knoweth to do good and doeth it not, to him it is sin."— James iv–17. Read and meditate.

DANIEL VENAMAN.

I will now expose an elder who lives in Macoupin county, Illinois, Daniel Venaman. He is a very active and prominent man in the Dunkard brotherhood, and is now out in Caliornia; I suppose partly on that mission, but I think more on a speculation for himself. He is amply rich in worldly trash, as his old colaborer, Johnny Metzker, one of Venaman's old cronies, says. They have traveled together a great deal, trying to do great work to be seen of men. They have solicited and begged enough money to buy a valuable piece of ground in the city of St. Louis, Mo., on which to erect a meeting house.

Over the vast brotherhood the interest was reported good for building up a church before the meeting house was erected, but soon after the house was built at a great cost, the interest began to decrease, and it is reported that at the present time there are only a few members there and no preacher. Their enterprise has been a failure, and there must be a cause for the failure. For one, we think the right man was not at the good work. It appears to me that the men who took such great interest in the enterprise were merely working it up to be seen and heard of through the papers, and to be elevated by men for doing some great act of church building. They made a great noise about it at the time, but since then they have taken action to have the house sold at the district meeting at Virden, Ill. There they decided that it was a complete failure, and concluded to sell the property. We think they have not builded their house on the solid rock, and it fell. Thus, if they don't change their evil ways, their spiritual house will be a worse failure than their St. Louis house.

Venaman is like Calvert and others; they are guilty of the blood of this persecuted family, because they helped to hide the mischief of their good old friend Metzker. No doubt but he will deny my assertions. He will certainly get himself deeper into the mire by denying the truth, for he is preaching in the pulpit that the truth shall make us free. He is not ignorant of the Scriptures' plain teachings, that a liar shall not enter the kingdom of heaven. And these are the kind of birds he was trying to save from being killed, and he went hand in hand with the old king bird. At the kings's call he was ready to obey him, and to do his wishes; he came and held a love feast in disguise, and under quiet deception he stood about in the same line of conduct towards me as did Jerry Calvert, and Calvert was caught in the untruths, and we think if Venaman will come up a few more times to Cerro Gordo at their bidding, to partake, with them liars, that he will likely become just as expert in telling lies as they are. They got so expert in that lying business that they have sworn to lies right in open court.

Some years ago I went to Venaman's house, and stayed all night with him. I laid my grievances before him, read a number of letters to him that were written to J. Metzker, and he seemed to hear me read the letters with some patience and sympathy. But, alas, he soon turned his coat, and not only that, but he gave me the cold shoulder. I then wrote him many letters, asking him to do his duty, and explaining my persecution in a way that he could not doubt the truth and believe a lie. I have referred the evangelist to many passages of Scripture to meditate upon, but he acted like one dumb and stupid, because he was bound to hide such heathenish acts. No doubt he believed, and does believe, all I said and wrote to him, for he dare not say I am a liar. But he had an object in view, and he would rather believe a combined set of liars than to believe me that old Johnny Metzker said, "Chris., I believe you are a Christian."

He is not willing to respond to the truths, and he took an active part to protect that cousining ring of Cerro Gordo. I have repeatedly written to him to try and save himself, and drop that stinking cousining ring. I warned him that he was even eating with those combined evil men and women. I told him that he should not encourage them to eat and drink unworthily, but he turned a deaf ear to my pleadings, and, of course, according to my understanding, he is as deep in the mire as are the cousining ring, because he rings in with them. I wonder what he will say before an intelligent community. Some persons, who belong to other denominations, have intimated that we have no intelligent people in the Dunkard Church; and some say, "Damn your Dunkards, we have seen enough of their supertitious ignorance in the court-room." Well, those are facts I do not deny. I say, let the sword cut whom it may. Right, and not might, will prevail.

I asked Venaman to send my letters back to me, but he did not do it. He wants to lay them before a committee. If the church does not investigate this act of persecution, then we might just as well say they are all ignoramuses, not worthy of notice. But I still have a better opinion of the Dunkard fraternity, and will contend for all that is honest and right according to the Divine Word of God and our Saviour's teachings. Venaman is the fifth wheel in this hypocritical vehicle at Cerro Gordo. He did himself injustice to save a cousining ring from the effects of a heathenish act of persecution.

I asked of him and his colaborers a permit to stand before their whole church to lay my grievances before them, and make my request known to the church. I have asked their official board to let me talk half an hour to their congregation, and then I would leave it to their consciences whether or not they should appoint an investigation to see if I stated the truth to them, but was denied the privilege to make my plea, and I think if Venaman had not been there I would have had my request granted, but I think he stood in the door with his broad horns of authority. He is a man of wonderful influence, and being wealthy he wants to be looked up to. I shall honor him for just what he is worth, not in dollars and cents, but in his hypocritical conduct. It is with him as with J. Metzker, and Calvert, and George Cripe, and some other humbugs. They preach humility, honesty, meekness, and equality in the pulpit, but they don't practice it. They can bind heavy burdens on others, and have a good deal to say about the evil of using tobacco and other various habits of human weakness. He is one of the kind that "strain at a gnat and swallow a camel." He might just as well try to swallow a camel, as to try and screen such heathenism from the effects of their dirty scrapes.

I want the good, thinking people of the Dunkard Church to mark that man, and not attempt to select him on a committee to settle this matter. I shall object to him for all time to come. Such a man as he cannot settle my business for me under any consideration. Col. Gibson, his colaborer, would have been willing to do something for this persecuted family; but, no, he was hushed up because he was not so high in authority, so he had a timidness of doing his full duty, as he wished to do. I have reason to believe that the colonel and his wife have suffered with this persecuted brother and his dear family. He answered some of my letters, and they showed sympathy, while others tried to see how cold and indifferent they could appear against me. A great many will deny this, but we must not forget that actions speak louder than words. The brethren who read this book will see at once that nothing short of a strong committee can save this Dunkard fraternity from being lost in almost heathenish darkness, but, if properly managed, it will have a tendency to purify and strengthen it.

Gibson answered a letter in this wise:

"*Dear Friend:*—In answer to your letter you have our sympathy in your troubles, but how to remedy them is the trouble with me now at this time, but I will seek counsel in this matter and let you know in the future. Stop lawing with them, for it is like one man lawing with a railroad company. They will break you up, and you will have to quit."

He surely knew that it was their aim to break me up.

A card from Mr. Eby:

"C. GIRL: *Dear Friend*—Your letter and card are at hand, but I decline taking any part in your case whatever, as I am too far away to do anything satisfactory, and I hope you will excuse me."

I will have to excuse him as I cannot do any better, only I will expose his conduct towards me afterwards. He had had a great many letters from me. He abused me shamefully at the annual conference several years ago at Darke county, Ohio. I went to him several times to ask questions, in order to get some secrets out of him, for I knew that he had heard some slanderous rumors about me, and I wanted to drive some things out, and I did drive something out, for I got him mad and then it came in rough style. I walked up pretty close to him, and asked him where my sister lived, as he knew. He said my sister was married to a Mr. Rhodes, and had moved to Southern California. Then I tried to show him how they had endeavored to overpower her when she was among the cousining ring at Cerro Gordo, by lying to her and by bringing all kinds of slanderous reports and lies against me, and even got her before a muddle-head of a Dunkard, and there she gave it as her opinion that her brother Chris. was insane. But he was not willing to hear my truths any further, and gave me a hunch and said, "I don't want to hear you, I don't believe you; I don't want to throw pearls before swine."

All that saved him from getting a slap in the face, was my slowness of anger, and the fear of arrest. I walked away before my anger came to me, and the farther I walked the madder I got. I took a circle around the ground and walked right back to the impertinent bishop, and I asked him whether he meant to call me a swine. He kept silent, and I told him that I demanded an answer, and he said, "I did not say you were a swine. I said I did not want to throw pearls before swine." He is one of the committee that was chosen on my first church trial. I heard old Johnny talked to them, as if he was telling them how to decide the case. What business had he go to and call them aside, and talk to that committee before they made their decision? He was the uncle of the heirs, and no doubt he got a percentage to intercede for them. Perhaps Eby does not like me because I brought the truth to bear on that committee. I don't owe him anything financially, but I owe him something religiously. I owe him good admonition, and will say to him to provide things honestly before God and man. Without these good gospel teachings it is impossible to please God. They undertook to screen an old king among them from being exposed to the world, and I must expose such combined devilment, or be separated from my dear family.

Just think, those men will go into the pulpit and preach that the truth must be told, and then they will try to hide the truth; aye, even do worse, they will lie in order to hide the truth. For they, being ignorant of God's righteousness and going about to establish their own righteousness, have not submitted themselves unto the righteousness of God, but to Israel he says, "All day long have I stretched forth my hand unto a disobedient and gainsaying people.

TRUST IN GOD.

Courage brother! do not stumble,
  Though thy path be dark as night;
There's a star to guide the humble—
  Trust in God and do the right.
Though the road be long and dreary,
  And the end be out of sight:
Foot it bravely, strong or weary—
  Trust in God and do the right.

Some will hate thee, some will love thee,
  Some will flatter, some will slight:
Cease from man, and look above thee;
  Trust in God and do the right.
Simple rule and safest guiding—
  Inward peace and shining light—
Star upon our path abiding—
  Trust in God and do the right.

SOLOMON SHIVELY.

I will not spare one who was a traitor and a deceitful outlaw, a son of the treacherous deacon, Shively, who has several times been expelled from the Dunkard Church. I shall give the names of all such traitors. He is called Solomon Shively. He is a chip off the old block. The outlaw seemed to sympathize with me for gain. He undertook to make me believe that he was a great friend of mine, and he helped me to make plans how to defeat my antagonists. He was good to me to get a chance to skin me out of fifty dollars. I had confidence enough in him to loan him that amount, thinking by his talk, that he had reformed his bad habits, and become an honest man; but alas, the last state of such a man is worse than the beginning. As soon as he had the money he began to turn the cold shoulder to me, and at my court trials he mingled right in with that cousining ring, and even got very familiar with them, and, especially in the court room, he would hang around them, and they were glad to have him assist them in their hellish persecution, and he was very anxious to testify.

In his conversation with me he called them a set of hypocrites, and dishonest. He came down on his grandfather, J. Metzker. He condemned even the whole fraternity, and he declared he would help me all he could. He promised to go out and talk to my son-in-law and my children. I knew that he had been put in jail for obtaining money under false pretences, but I was like a drowning man in my persecution, catching at straws. But he was worse than a straw. He was what I would call a turncoat and a rebel. He was like J. M. Rainey. He favored them, and of all the devilish testimony produced, he laid all the rest of the cousining ring in the shade. He beat them all. I think he was coaxed to come and swear that he would not believe me under oath, but Lawyer Jones has pretty nearly used him up. He exposed his devilment, and they came near impeaching his testimony, and we think if he had got his just dues, he would not have escaped the penitentiary. But he did the cousining ring a great favor. We think they got him to do the rough swearing to help them hide their mischief. Before, he would not mingle with them, and they also did not want anything to do with him. Since those law suits, they have become very sociable together, and he moved among them at Cerro Gordo. I have heard some talk that they will take him into the church again. We think that is the place for him, right in that cousining ring. He can do some good swearing for them when they get in trouble. I hope they will put him in office, for they owe him some honor for giving them such a big lift with his evidence so nicely in their favor. I wrote him a letter, and told him of his treacherous actions, and of his meanness in betraying me for self gain. I think I told him that he was as ornery as a dog for betraying a persecuted man. He sent me the following lines:

"DECATUR, ILL., June 28, 1885.

"MR. C. GIRL: Sir—You certainly missed your aim when you tried to scare me. As for doing like a dog, the least you say the longer you will be able to lie down. I would not stoop so low as to strike you, but I will turn you over my knee and slap you till you whine like a hound pup. Now make no more threats but go to work like a man. As for trouble, I assure you, you will get your share."

Well, as for trouble, he had it in his head to make me trouble, to stand in with such a set of rascals to give in such evidence. It is enough to make the best saint tremble to hear it fall from the lips of an outlaw and a treacherous robber.

The time myself and daughter Ida took a trip to Florida, in the year 1885, he got the fifty dollars. I also gave him a note of four dollars and fifty cents for collection, and the man told me that he offered him the note if he paid him, I think he said, one dollar. He was hard up all the time for money, and he came to me at various times asking for just one dollar. He got, perhaps, at various times, three or four dollars to buy provisions for his family. I think he gambled some of it away instead of paying his grocery bills. Brock Deardorff told me that Solomon said they could get some money out of the old man anyhow, and he was right about that. They succeeded, but they did not obtain it fair and honestly, and it will be a hindrance to their future prosperity. That is what Deardorff told me, and the people in Decatur know that Deardorff would not suffer himself to be caught in a lie. He would rather tell some more, and try to get out of one in that way. Old Bishop Wagner says, "that a man that lied once would lie again when he was pinched."

The trouble with Solly Shively is, he was not raised right. His parents showed too much deception in their christian professors. They were worshipping the critter instead of the Creator, and there are a good many more people in and about that cousining ring who are in the same way, and they have surely run their gospel ship aground. It is most astonishing, that with all the intelligence and the many christian privileges we have in this nineteenth century, to see intelligent men and women worshipping their preachers instead of their Creator. My dear fellow citizens, we have to look up higher than man to be saved. What is poor mortal man? Let us wake up to our christian duty:—first to our moral duty, and then we have only a step or two to make to become Christians. "He that hath ears, let him hear," and "harden not your hearts," said He unto His disciples. It is impossible but that offences will come, but woe unto him by whom they cometh. "It were better for him if a millstone were hanged about his neck and that he were drowned in the depth of the sea."

A man would be in a nice fix if he was cast into the sea, or even into a fish pond, and then told to swim or drown. The Lord will help us swim if we will make the effort according to our Saviour's teachings. He is ever near; never fear. He will help in all times of need. Don't forget that, my dear reader. Knowing this, the trying of your faith worketh patiently. May God help the hypocrites to turn from their evil ways. Fellow citizens, let us look to our eternal interest, and be blest forever.

"We may write our names in books, we may trace them in the sand,
We may chisel them in marble with a firm and skillful hand,
But, my friends, there is a book, filled with leaves of purest white,
Where no names are ever sullied, but are ever pure and bright."

There are many names written; may they all be written in the book of life.

BISHOP CALVERT.

I will now expose another bishop, Mr. Calvert, who was brought to Cerro Gordo from the State of Indiana. I have known him for many years. A few years ago he was called to Cerro Gordo to hold a revival meeting. He is what the Scriptures call a hireling. We understood they paid him two dollars a day to preach for them. He did some pretty loud hollering and blowing and bragging, so much so that the people got tired of hearing him; even some of his own members got disgusted with him. As soon as I heard that Jessy Calvert was up in Cerro Gordo I wrote him a letter, asking him to come to Decatur to see us, and telling him that I had some important business with him. I told him the official board of the Cerro Gordo Church was very much out of order, and the first thing to do would be to get that board in order with some of its lay members, and then hold a good revival meeting, otherwise he could do no good. I told him that unless that was done that it would be merely mocking; but Calvert took no heed of my good and wholesome counsel, but he slashed away careless, and made himself bold in condemning other denominations, and he soon got into deep trouble with the Methodist and other churches. It seemed as though he tried to condemn everything but the Dunkards. He soon got caught in lies, and they denounced him a public liar, and proved it in the Illiopolis papers, but he went on to try and convert souls to God.

He would have likely got a short reproof for not answering my letters, but he would not let me get near him. He seemed to shy away from me as though I was a pickpocket or a hypocrite. It seemed as though I was not to touch the hem of his garment. I came home very much discouraged, thinking to myself, here is another treacherous devil to fight. I saw the very mischief and evil spirit in that man. I came home and opened the Bible, and it seemed that the first chapter my eyes met was encouraging for a persecuted man. I would commence to fight a new devil, let him present himself in any shape. I could find in one chapter of the Bible more encouragement than all the preachers could or would give me. I wrote another letter to Calvert and copied it in order to expose him. With his loud talk and excitement to try and scare people to join the Dunkard's Church, he succeeded in getting a few young converts. He blowed around there for several weeks, and then he went, slinking away like a thief in the night. His treacherous actions gave him to understand that I would expose him in my history. We will publish some of my admonitions to him in the latter part of this book.

I went the second time to see him, and happened to meet him on the sidewalk. He was alone, going to the meeting house to fill an appointment. He was coming from the east, and I was coming from the north to meet him at the crossing, but he passed by ahead of me. I hailed to him but he did not stop. I hailed to him the second time, and then he stopped, but not long enought to have a talk. I reached out my hand and said, "Calvert, do you know me?" He said, no, he did not. I introduced myself to him in this way, "I am this persecuted Chris. Girl." "O, yes, I have heard of you," and he walked on. He would not stop to talk with me. In front of the meeting house stood the old king of the cousining ring and his crowd.

I told them that they were counseling for devilment; that "birds of a feather flock together."

See Ephesians, fifth chapter, sixth and seventh verses:

" Let no man deceive you with vain words; for because of these things cometh the wrath of God upon the children of disobedience.

" And have no fellowship with the unfruitful works of darkness, but rather reprove them."

Also, II Timothy, chapter third, verses first and second:

" This know also, that in the last days perilous times shall come.

" For men shall be lovers of their own selves, covetous, boasters, proud, blasphemous, disobedient to parents, unthankful, unholy."

Here the reader will see that God will not suffer us to be persecuted more than we are able to bear. Oh! the evil and devilment there is in those bishops, who are standing in the pulpit, assuring the public that God is an allwise being, and knows the very intent of the human heart. It does seem that God had sent them a delusion, and made them believe a lie and be damned. Shall we, as a christian people, suffer the devil to come behind the pulpit to tell us the way to heaven? Have we not enought before us in print to make us wise unto salvation?

Mr. Calvert and I met after he was caught in the untruths. We went to a meeting in Darke county, Ohio. I told him, before the editors of "*The Gospel Messenger,*" the Dunkard paper, that he was caught in a lie out in Cerro Gordo, by a Methodist preacher, Mr. Reasnor. He made no reply, but walked away, knowing that the truth was staring him in the face. He could not bear the pressure and sneaked off. It was quite an undertaking for me to boldly expose him as a liar before those editors. My dear readers, it was all I could do to muster up courage to reprove him for his former conduct. It seemed a necessity on my part to stand up to the work that was before me. I would have rather stepped up to that old bishop and said, "How do you do, Brother Calvert," and had a social talk with him; but he denied me and said he did not know me. It was an untruth, for I had known him ten or twelve years. In order to excuse myself to those editors, I said, "Brothers, to prove to you the assertions I have made against Calvert are true, you can read this," and I handed them the papers containing the proof. And still that man had the boldness to come back to Cerro Gordo to hold a revival meeting, the next fall, but a majority of the better thinking members would not go to hear him; he was looked upon as a humbug. As he only got the support from that cousining ring there was no life in their meetings. A drag and a dead weight of sin hung about them, because they have transgressed the laws of God, and have eaten and drunken unworthily to hide their deception from the eyes of the world. Oh, what mockery I have seen for many years. May God be merciful to the hypocrites and turn them from their evil ways. There are many who are worshipping idols instead of their Creator. I hope and trust that this volume may be the means of turning many to examine themselves and see whether they are serving the true and living God, and not man, who is made of the dust of the earth. What I have said about Mr. Calvert is the truth, and will stand when heaven and earth shall pass away. I hope, therefore, that he and many others may accept the truth and repent from their evil ways before it is forever too late. Remember that God is not to be mocked. What we sow, that shall we reap. I have no malice toward him; God forbid.

ELDER MILLER.

I will now proceed to speak of Robert Miller. He has had ample warning years ago from me concerning his exposure in my history. I will not spare one, not because I hate them, but for the love and good that is left in the church, and for our admonition to be waked up to our Christian duty before it is forever too late. This thing of pleading ignorance as an excuse for a failure to perform our christian duty, will not do, for the Saviour has made it so plain that "a way-faring man, though a fool, may not err therein." How, then, shall a man err in such a plain case of persecution as this before us. How a lawyer and a preacher like Robert Miller, who lives in North Manchester, Indiana, one who studied law, and I suppose, practiced in the civil courts of justice, but who has abandoned the practice and become a Dunkard preacher, and a very active minister in advocating the cause of

Christ and his kingdom, — how, I say, he can shrink from his duty in such a plain case of persecution as this, is more than I can understand. I will refer him and others to the tenth chapter of I Corinthians:

> 1. Moreover, brethren, I would not that ye should be ignorant, how that all our fathers were under the cloud, and all passed through the sea.
>
> 5. But with many of them God was not well pleased: for they were overthrown in the wilderness.
>
> 9. Neither let us tempt Christ, as some of them also tempted, and were destroyed of serpents.
>
> 11. Now all these things happened unto them for ensamples; and they are written for our admonition, upon whom the ends of the world are come.
>
> 12. Wherefore let him that thinketh he standeth take heed lest he fall.

Brethren these simple passages have a meaning in them.

Well, we will give the facts in this particular case. We will show that there was a great neglect of duty, and the Scripture tells us that "he who knoweth to do good, and doeth it not, to him it is sin." I have drank of the bitter cup of persecution for twelve or thirteen years. I have been witness of the unfair ruling from time to time, and it has gone on from bad to worse. I tried to correct it by moral suasion, and, as I thought, according to the rules laid down in the Blessed Word of God to the best of my understanding, and according to the Dunkard discipline, but was overpowered by the many evil doers, who deviated from the plain commandments laid down for our instruction and our correction. Many of the brethren whom I called upon pretended they were ignorant of the persecution which has been going on for the last twelve or thirteen years, and I will be very free to name them.

I will now give the proceedings of the first trouble I ever had with the official board of the Cerro Gordo Church. I bought a farm of one of the old preachers. In giving the notes one of them was written by mistake at ten instead of six per cent., which made a difference of eighty dollars at the end of four years. The mistake was overlooked, perhaps, by both parties. When I discovered the mistake I asked two of the heirs what we would do about it. They told me to go and pay it, and they would pay the eighty dollars back to me. I paid the note in full, but the heirs refused to pay back the eighty dollars. They were two deacons in the church. It is also as well to mention that they were nephews of old Johnny Metzker, and were of the elect official board of the church. Well, as they refused to do what they had promised I brought the matter before the church, and the matter came before the council for investigation. The official board stuck their heads together and proposed to me that the matter should be settled by a committee of three strong brethren. I objected to their plan for various reasons, but some of the strange brethren were sent to me to persuade me to yield to their plan of leaving it in a committee's hand. After much persuasion I yielded to their wishes, and the following brethern were appointed: Henry Davy, of Ohio, Enoch Ely, of Stephenson county, Illinois, and R. H. Miller, of North Manchester, Indiana. Old Johnny Metzker had a good deal to say to them before they retired to decide this important matter. They decided against me, and I lost my eighty dollars, although I proved to the committee that the bishop had sold his land to me at six per cent. interest on the notes. The heirs acknowledged that they had agreed to pay the eighty dollars back, but they added an "if" to it. They said, "We said 'if' we owe it." At the time they made the promise to me they had no "if" in it, but said, "You pay it to the executor, and we will pay it back to you." I contend that the committee had been induced in some way, perhaps not willfully, but by some wrong influence, to decide wrong in that matter, and I suppose they would not dare to say that they could judge properly every time. The wrong they have done is not in giving a wrong decision, but in letting themselves be overpowered by evil influences against their better judgments, and thus they are carried along to this day, and are giving their influence on Satan's side; and by so doing they are tearing down instead of building up. In trying to save one Judas from being exposed, they may drive and keep thousands out of the church; and it is astonishing to see a man like Robert Miller yield to such an influence. He has always treated me with respect when we have met, but I have a complaint to make against him, not to injure him, but to instruct him, and if he is as willing to take counsel as he is to give it, we think he will take the admonition in a christian way, together with the reproof and warning we have given him from time to time.

I met him in Darke county, Ohio, at the annual conference. I gave him rather a sharp reproof for not answering my many letters of appeal for help. Mr. Miller tried to make a satisfactory apology, but did not succeed. He said: "I told my wife that I ought to answer Girl's letters." Ought to! Is that binding? According to his own preaching it is binding. I was willing to let him off with his apology, but I made a renewed request, and even sent him stamps to send my letters back to me to use them in my history, but he sent neither stamps nor letters. We wonder if he pays his debts with promises and apologies. I guess he will say: "I ought to correspond with that persecuted brother, but I did not do it; and I ought to be impartial but I was overpowered by my colaborers to be willing to favor them in their sinful ways, but I will make that all right in making apologies when the time comes to do so."

I know who I am writing to, and I will not make my remarks very lengthy, but will say to R. H. Miller, and whom it may concern, that God is no respector of persons. Please turn to the twelfth chapter of I Corinthians:

"Now concerning spiritual gifts, brethren, I would not have you ignorant.

"That there should be no schism in the body; but that the members should have the same care one for another."

JACOB REPLOGLE.

Jacob Replogle is a man that had lots of trouble in the Dunkard Church. He had many trials before the official board, and also the different precincts of the church, and he had a good many before old Johnny Metzker, and he always came out beat and dissatisfied, and complained bitterly about the unfairness in their ruling in church matters. He said that Johnny Metzker was not fit to rule a a church; that old Johnny was not childish, but devilish, and he called Jacob Wagner hard names. He was formerly a deacon of the Okaw Church, and he had a great many hard things to say against the official board of that church.

I happened to have some dealings with him, and that is by far the best way to find out what is in a man or a woman. I bought a house and lot of him in the town of Cerro Gordo, and he agreed to furnish me a new abstract in case I wanted to sell the property. He gave me an old abstract that had been made out with some other lots before. They were separated a half block, and he came to get the abstract to see about them. He kept the abstract and went off to Kansas, and while he was gone Mr. Clifton made me an offer for the house and lot. I think old Johnny Metzker put him up to buy me out. I was close to him, and he wanted to get me away as he did not want to live close neighbor to a christian. Well, I had to have an abstract or lose my sale, as Clifton said he would not take the property unless he had a new abstract, which Replogle agreed to furnish as soon as it was needed, but he refused to do as he had promised, and then denied that he had agreed to do it. There being so many transfers to make out in the abstract it cost over thirteen dollars. I thought that if he would not pay what he agreed to I would sue him. He told me to crack ahead, and he beat me on the same ground as the cousining ring, with bad evidence. There is no use for me to start a law suit with any of that cousining ring, as they will ring me out every time. They can swear harder under oath than what my conscience will allow me to, and still they want to be careful that they don't break their baptismal vow, not to swear at all, but to affirm, which is just as binding as to take an oath. Which is the worst: to swear to a lie, or affirm to a lie? Some will say it is as broad as it is long, but there is a good deal of difference in affirming to a lie and swearing to the truth. May God forgive them as I am willing to forgive them. It is getting to be fashionable to get on the witness stand and rattle off a whole lot of lies before a judge and jury. I am treating of facts that cannot be denied by any good, honest man without putting a mark of dishonesty on his character. Our American people will have to do better in moral principles, as well as in christian principles, or they will be laid in the shade by some of the so-called heathens who have been worshipping idols. They are improving while we are losing ground. He that hath ears let him hear and understand.

WILLIAM GIRL AND OTHERS.

I will now give to the reader a transaction I had with my son William. I at one time borrowed some money of him, and gave him my note the same as to a stranger. Now to show that my children were under the influence of those ungodly men. When that boy wanted his money I told him to come out and tell me what those men were trying to put him against his parents for, and for what purpose, and I would pay him his money, but he would not tell on them, and got mad, and threatened to sue me on the note. I told him to sue me, and then I would bring in an offset against him as he was of age, as I had boarded him and a horse for about five months each winter for three winters. But he chose the opposite side rather than be obedient to his parents, and he went to 'Squire Nickey, that old flat-headed Dutchman, and he gave him counsel, and not knowing enough about law to know whether I could bring an offset against my son, he goes to another 'Squire for advice in the matter, who told him he did not know whether he could or not, and then Nickey went home and planned it for my son to let his (Nickey's) son have the note for collection, and thus probably avoid the offset; but before they sued the note I offered to pay my son, when he told me he did not have the note, that 'Squire Nickey's son had it. I told him to go and get it, for I did not want to go there for it, as they had treated me mean and disrespectfully already, and it was my wish to avoid them; but no, he would not make any attempt to get the note or take the money. I and my other son were going on a visit to Ohio, and they waited until the day we started on our visit, and then sued me, and by the time I returned they had the suit, costs, and all against me, and I paid it like a man. Why did they not sue me when I was here to defend myself? It must have been because they thought the way they planned it was cunning and smart, but it showed their hated, vicious, ignorant cowardly conduct against me. It appears that the more sinful and wicked they are, the more religion they claimed to have. How can a man think otherwise when they will go and confess God in worship and in prayer, and then go and participate in such hellish, dartardly actions as this. There is no religion in them; they only use that for a cloak to hide their hellish crimes, for it is in the very marrow of their bones to be hypocrites, and it takes more than feet washing to cleanse them. They would have to be put in a vat and boiled and soaked, and then put through a purifying process for about a year before they would be agreeable to man and society.

Now, remember, that God in His infinite mercy says, repent and believe and you shall be saved. Dear brethren, you cannot be saved by mockery, which you practice, for God says, "I am not to be mocked." I have plead for my children to discard all evil influences, and take my good, fatherly counsel, and I assured them that everything would be for their good. I wanted them to be good and happy children, and never be the cause of tears trickling down their parents' cheeks. I hope, by the grace of God, they may finally be saved.

SIMON NICKEY.

I will now begin this glad new year of our Lord Jesus Christ, 1888, with an account of the rottenness of that cousining ring of Cerro Gordo, Ill., by giving the testimony of Simon Nickey against me. It is an evident fact that Simon Nickey was selected as my guardian in case they succeeded in getting me off to the insane asylum. Of course he denied the assertion to me. He has taken active part with that ring of conspirators in helping to prosecute me, and I was told that 'Squire Nickey went to Mr. Sowarn and tried to persuade him to go to town with him, and see whether they could not accomplish the feat of sending me to the asylum, but Sowarn refused, for he well knew that I was competent to take care of my little property, and the Irishman did wisely to keep out of that unfair and treacherous piece of business.

The 'Squire always tried to make me believe that he stood neutral in this conspiracy. I went to his house one Sunday morning to have a talk with my son William. I tried for a long time to get the secret conspiracy out of that son, but all in vain. I tried every possible way to induce him to come out and tell on Bingaman, C. Funk and a few others, but he was forbidden to tell on them. Finally the boy went away from me and started for the house. We were in the barn. Simon came out and I had a long talk with him, laid my grievances before him, and asked him many questions concerning the matter, and he questioned me. I answered him promptly, but he was slow in answering me. In the winding up of our conversation I told Mr. Nickey that he had a way to get those secrets out of the boy, and he asked me in what way. I told him to mob him and scare them out of him. "Why, you wouldn't hang or hurt him?" said he. I said no, only scare him, and Nickey testified that I only wanted to scare him. I told the same thing to Mr. Phillips.

They arrested me the second time, and put me in prison. The first arrest was made July 17, 1883, and the second arrest was caused by my confiding my plans of scaring the boy to S. Nickey and J. Phillips, supposing them to be confidential friends.

I have often visited J. Phillips and have explained my troubles to him and his wife, and they seemed to sympathize deeply with me and my family. I read several letters to them that I wrote to the official board of the Cerro Gordo Church, but they betrayed me and became workers against me. As soon as I had laid my plans to try and scare the secret out my son, J. Phillips went and informed him of the threat I made, and that started the boy's malice afresh, and by counselors, such as 'Squire Nickey, A. Bingaman and C. Funk, he was induced to have the second arrest made, and had me placed in jail. It was only by the persuasion of such evil men that it was done, for Wolgamot and others testified that the boy told them that he had never done anything in his life that he hated so bad as to have his father arrested. Mr. Wolgamot asked him who urged him to do it, and he said, Funk, Bingaman, and others.

One day my son and C. Funk passed my house, I said to them, in a joking way, "are you going out to find a guardian for me," and they laughed, and C. Funk said: "That is what we are after." Sure enough that was their errand, and soon after the arrest followed by Sheriff Wetzel at my stable, where I was with my hired hands getting ready to go and make hay. The sheriff testified that I gave my hands orders how to do my work while I was absent. I then told him I was ready. I stepped into the buggy with the sheriff, he brought me to town, and I was landed in jail to await my trial. Sheriff Foster not being at home the deputy sheriff had me put in jail by the sheriff's orders.

In a few days Sheriff Foster came home, and came into my room, conversed with me, and said " Mr. Girl, it is rather hot in this room." The sun was shining in from the west through two large windows. He took me out of that room, and escorted me to a well furnished room down stairs, and told me to make myself at home. I laid down on the sofa to take a good rest, and at meal time I was invited to sit up to the family table.

May God help us to turn from darkness into light, and not worship man, but the living light, is my prayer. It is not hard for the christian man to bear his burdens when his heart is washed in the blood of the Lamb. His christian hope keeps him strong in the faith of the Lord.

I have contended with my fellow men in the church and family troubles with a zealous religious feeling, and have tried to avoid all strife and malice, and exhorted everyone to do the same, but it seems as if the evil spirit's influence was on them, and that they courted and worshipped the devil instead of the divine truths of God, and it compels me to lay the facts before the people in the shape of a history, that all good thinking people may judge for themselves.

I will now expose to the world a lot of professing christian men in their hellish work, such as elders and deacons of the brotherhood. I will proceed with care to give the details correct, as they happened, endeavoring to lay the truth before the public, as we can read in the blessed Word of God, that the truth shall make us free, and we shall be free indeed. We American-born people want to sustain our rights, especially in religion. We want to be sure that we are right according to the Saviour's instructions, and then drive ahead although the devil stands at the door. The Rev. Sam Jones, in one of his sermons, says, "if a thing is right, fight for it, and fight on though you may feel sometimes that you are alone; but fight on, and when the battle is over and the smoke is cleared away, you will see God and His angels, and good men and women standing around to cheer you."

Many have started in this religious warfare with such a determination, but soon something comes in their way, and they begin to murmur that the cross is getting too heavy to bear, and they get weak in their faith. They forget that Christ will help them to bear their cross, and that he will be with his people in their times of need. A backsliding professor is, in my opinion, a great hindrance to the true principles of the christian religion. I once heard a pastor in Palatka, Florida, say in his sermon: "My friends, I don't want you to hold out, but hold on to God's promises, and he will finally come to your aid and fulfill his blessed promises. He says the believer in Christ shall not be tempted above what he is able to bear. Let us look to God for our help. He is ever near."

"There is an eye that never sleeps beneath the wing of night;
There is an ear that never shuts when sink the beams of night;
There is an arm that never tires when human strength gives way;
There is a love that never fails when earthly loves decay."

REV. G. W. WILSON.

I will not confine myself entirely to Dunkards, as I have been badly treated by others. I will mention the Rev. G. W. Wilson, an evangelist, who presided over a revival of about three months' duration in Stap's chapel, at Decatur, Ill., in the winter of 1887. He seemed to be a well-read man, and very ambitious in our Master's cause, and did all he could in the pulpit to convert sinners to the cause of Christ. I enjoyed his sermons at first and approved his work, for he labored hard to warm up those who became cold in the cause and put them in working condition. He did not spare them, and spoke very pointedly to them. He seemed to follow Paul's advice in II Timothy, fourth chapter, second and third verses :

"Preach the word; be instant in season and out of season; reprove, rebuke, exhort with all long-suffering and doctrine.

"For the time will come when they will not endure sound doctrine; but after their own lusts shall they heap to themselves teachers; having itching ears."

I think he took Paul's advice, for he preached the word to saint and sinner. He created a great interest among the members, and they became so happy they could not contain themselves, and they gave praises to God. I felt that Mr. Wilson was a good christian, and I thought it would be of much strength to me to have a talk with him, and pour out my troubled soul to him. I thought he could give me good advice, and I asked a friend to intercede for me. He did so, and I asked him to come and see me at my office, and have a christianlike talk with me. I thought if he wanted to do so much good he would surely come and talk with a man in great distress. At first he gave me great encouragement about coming to see me, but finally he gave me the cold shoulder. I wrote him a letter requesting him to come or write an answer, but he paid no attention to it. I then dropped him a card and told him to either return my letter through the post office or come in person, as I was writing a history and wanted to put the letter in it.

The letter and card soon came through the post office. I wrote him again, explaining my persecution, and told him that he should answer or come and see me at my office. He made no effort to do so. I have sent him several propositions to come and see me, and save himself from being blasted, as his actions showed that he was not going to stoop so low as to talk Scripture to such a plain-dressed old farmer, an expelled Dunkard, and we think some of the members of his church sided in with the Dunkard clique, and we think they put a bug in the evangelist's ear, and it was a humbug; so the reverend gentleman got humbugged by some of his own members, and he showed by his actions that he would have nothing to do with this persecuted Dunkard, but instead of assisting me with advice and counsel, he waited until I was absent from the city and then made light of my troubles.

I think it showed a most cowardly spirit to make sport of me, after I gave him several texts from the Bible to preach from. Instead of selecting my texts for his sermons he made remarks about me. I don't know, and don't care, what those remarks were. Of course his audience knew, and they ought to know that he took a sneaking way for doing it, as to wait until I was absent. Some of his members got disgusted with the foolish remarks he made.

14

He first said that the Methodist Church was asleep, and needed an awakening. Then, when he he got them a little excited, commenced making his brags that the Methodist was the best church in the universe, saying they had three millions of members. It was well that he did not say they had three millions of *christians* in it. We hope, when this strikes Mr. Wilson, he will hereafter sit down and counsel with a man or a Girl when they make an earnest appeal for help as I did.

I told Mr. Wilson that if I had to paddle my own canoe, then their preaching was all in vain;— that it was a dead expense to support ministers by paying them big salaries to stand in the pulpit to confuse the people. If we can be saved by the Divine Word of God, why not save those big salaries and give them to the poor; and if the preachers kick against that, send them off to the insane asylum or to the penitentiary.

What we need in this day and age of confusion, is the conversion of skeptic preachers. I wrote to the Rev. Mr. Wilson, that if he would convert all the skeptic preachers, I would bring up the rear, and would convert the world. He undoubtedly made light of that, and put me down as a crank. I would rather be called a fool for Christ's sake, than be fooled by a skeptic preacher who stands in the pulpit to lead his fellow men and women to destruction. Let us be careful and look out on that line. I honestly believe that many church members worship their pastor more than their Creator.

Copy of a Letter to Rev. Mr. Wilson.

"*Dear Brother Wilson:* We sometimes say the third time is the charm. I have asked you, I think, three times to have a christian talk with me, but your actions show to me that you will not do it, and I cannot force it upon you. I can tell by your actions, for actions speak louder than words. You spoke very loud last night, and as far as I could comprehend your words, they were very much to the point.

" Mr. Wilson, I like your way of preaching — you are so personal to the professing Christian. That is right. You cannot hammer too hard in that direction. It does my soul good to see an Evangelist hammer at the very root of the most damnable evil that the American people are guilty of; that is, to join a church for popularity, or to be honored by men in order to move in the society of aristocrats, and for self-gain in accumulating the almighty dollar. Oh! how can the love of Christ be in the hearts of such church members? It is easy to be a professor of christianity, but it is much easier to be an oppressor. If you once get the love of Christ in your heart, and have tasted of His great love, then the work is done. We cannot help being Christians. If God is for us, who can be against us? The very gates of hell shall not prevail against us. Though our lot is cast among demons, hypocrites and backbiters, devils, or even the devil's angels, shall not prevail against those who firmly trust in Our Redeemer. We shall not be persecuted above what we are able to bear, if we are planted firmly on that rock. May God help you as an Evangelist to make the good people realize that we must get down on the solid bed-rock, and our house will stand for evermore. Amen."

A RECOMMENDATION.

MUSCATINE, IOWA, June 27, 1885.

Dear Old Friend: I will try to answer your last letter. I got the check for $10.00, and am thankful for the same. As to how you acted while here at my house, I cannot say anything but that you was a gentleman in every respect, and my wife says the same, as do the rest of the folks, except George and his wife. I don't know what they say; that is for you and them. I am sorry your wife is sick, but hope she is better by this time. I am in a hurry, and have not time to do you justice at present. I will send your note, and will try to do better next time.

Yours in Love,

T. G. THOMPSON.

A LETTER.

CERRO GORDO, ILL., Sept. 4, 1887.

TO THE EDITOR OF *The Gospel Messenger*,

Mr. Editor: I will not call you brother, for you do not act like one towards me, and you have now shown your colors as plain as there is any need. You had better stop now and consider what you have done. For the last three or four years you have tried your very best to shield these hypocrites. I have warned you that you would expose yourself, but have gone on with your cowardly acts and tried to screen a number of your colaborers from a hellish and heathenish work. Instead of answering my questions, which I asked you to publish, so as to give the whole brotherhood a chance to reply, I suppose you have thrown them in the waste basket, with other good letters and good articles, which, although you admitted they were good letters and good articles, you called them anonymous, and made that your excuse for not giving them room in your paper.

It seems as though you have no room in your paper for my name; you will have it on the wrapper, but I think you would rather not have it even on the outside. I have found that the charities asked for by me are growing more frequent, and I have found, to my sorrow, that in some cases the parties who called for help were not deserving. When help is needed it must be sought for in an authorized way. Now, Mr. Brumbouck, you, as an editor of the *Gospel Messenger*, are holding a high position, and you should not show partiality and try to favor such as make and believe lies. You may try to frame some excuse or apology after you are caught in the trap with some of those unruly devils. I have no better name for them until they stop their devilish acts and repent of their evil ways, and if you will advocate their heathenish case and try to hide these ungodly men, you are no better than they are.

Those several articles which you refused to publish in your paper will stare you in the face in my history, and you cannot help it. Neither can I help it that you tolerate such things against me, a persecuted one, helpless from most all human aid, and that you are casting it on me as severe as you can. Can you not see that you are making haste to ruin the whole brotherhood of Dunkards? I am now separated from a once-loved family, and all through the influence of those that I have mentioned to you long ago. I am filling my history with numerous letters like this one. I suppose you want a copy of it, which I will try to finish before long. Please send in your order at once in order to secure a copy, as there will be a great demand for the history of the supposed crazy Christian Girl. That is my name. The old veteran soldier, as some of you call him, has declared to my face that he believed I was a christian, and believed I was trying to do right before God and man; but the day previous he said I must go to Jacksonville if I did not look out. I have been watching along the line of that cousining ring. The old money-god is king of the gang of heathen outlaws — not only outlaws in court here on earth, but they have violated God's laws where God himself is the Supreme Judge, and not Editor Brumbouck, or any other red-headed editor or preacher. I suppose red hair is honorable if the man that wears it is honorable. I had business with several red-haired men in the town of Cerro Gordo and they have treated me very mean, one in particular that held the office of marshal. He arrested me for disturbing the peace. He was the cause of my being fined fifteen dollars and costs, just for spite. He is a renegade Dunkard, and says he belongs to what is called the Christian Church. He arrested me because I called him a renegade Dunkard. He flew all to pieces, and tried to jerk me off my buggy, and took my cane away from me. He caught me by the arm, and led me on the sidewalk down to the 'Squire's office. A big crowd followed us. He kept abusing me as he led me along, and said I was drunk. I told him that he was a notorious liar. I have lived in this neighborhood for twenty-two years, and no one has ever seen me drunk. You ask old Johnny Metzker whether I am a drunkard or not. They have some such in J. Metzker's church. The marshal is in the employ of a rich Dunkard, and he helps them to fight for self-interest. Some of you Dunkards are all the time condemning tobacco, and my observations are that the most ungodly and the most unprincipled men are the ones that are hammering on the tobacco question. Such that would not choke at a lie, and those who have sworn to lies, I have heard talk loudest about the evils of tobacco. I heard an old bishop say or intimate that a man could not be a christian and use tobacco. He talked as though he had his whole mind on the filthy weed, and yet at the time he talked he was after a rich widow sister, and his wife had been buried but a few months. He overheard me say to a good brother friend that she would make a good wife, but the old bishop smelt a rat, hurried up the cakes, and got the rich widow.

T. J. Rosenbarger, who was here early in the spring, preached and hollered so that he could be heard all over town until his lungs got sore; but went home without any additions to the church. Sane and solid-minded men and women will not join themselves to such a corrupt cousining ring. A deacon was dangerously sick with typhoid fever. I understand that he is lamenting, and is afraid if he dies he will be lost. I understand his sickness. He is a sin-sick soul, and I have written out a prescription for him, and for the whole cousining ring; but the trouble is they don't take the medicine, and they will grow sicker and sicker, and finally die and be damned to hell. Not so, because I say so, or you as a preacher in the pulpit say so, but we understand the Scriptures that way, and they are plain and easy to be understood. What shall become of the hypocrites, and liars, and drunkards? They shall not enter the kingdom of heaven.

And now, Mr. Editor, if my writing don't digest well with you, get down on your knees and ask God to take that stubborn stone away, and pray God to help you pull the beam of partiality out of your eyes, and then thou shalt see clearly how to pull the splinter out of thy brother's eye. Then God will make a way possible for you to escape the snares of the devil. I am no editor, nor preacher, but I am sure that I have been persecuted even to false imprisonment, not by my eldest son William, but through the cousining ring's influence. My family had not wisdom enough to do such mean work without the devil's evidence. Bear in mind that this is evidence of their evil deeds, and your silence proves that you are in the mud as deep as they are in the mire, just to screen these men.

Yours truly,

CHRISTIAN GIRL.

SCANDAL LANE.

It is not on the signboard, sir;
   Go search both far and wide,
Or in the town directory,
   The map or railroad guide.
And if you pump your neighbor, sir,
   You pump, alas, in vain,
For no one e'er acknowledged yet
   He lived in Scandal Lane.

It is a fearful neighborhood,
   So secret and so sly,
Although the tenants oftentimes
   Include the rich and high.
I'm told they're even cannibals,
   And when they dine and sup,
By way of change they'll turn about
   And eat each other up.

They much prefer the youthful, sir,
   The beautiful and rare;
They grind up character and all,
   And call it wholesome fare.

And should the helpless victim wince,
   They heed no cries of pain,
Those very bloody cannibals
   That live in Scandal Lane.

If you should chance to dine with them,
   Pray never be deceived,
When they seem most like bosom friends,
   They're least to be believed.
Their claws are sheathed in velvet, sir,
   Their teeth are hid by smiles,
And woe betide the innocent
   Who falls beneath their wiles.

When they have singled out their prey
   They make a cat-like spring,
Or hug them like a serpent ere
   They plant the fatal sting.
And then they wash their guilty hands
   But don't efface the stain,
The very greedy cannibals
   That live in Scandal Lane.

PLEADING TO EDITORS.

I have plead with the editors to publish my articles that I have written calling for a committee to investigate my trouble, but the editors would not heed my call, because it would expose the ring of ungodly men in their acts of breaking up my home and family. They would rather screen those bishops and elders, as they contributed to the support of their papers, and a dollar in their eyes looks as big as a wagon wheel. How can so-called christians become so depraved as to forsake all religion, and help to crush a family for a few cents.

Here is a copy of a letter I wrote to the editor at Cerro Gordo:

"Mr. Editor — Please to make known my plea and my wishes through your paper to my fellow citizens, to show how plain is the conflict between right and wrong; that is, give me an equal chance with my opponents, and allow me to lay the truth before the public. There have been a great many mysterious things come before the court, by the way of evidence, that need an immediate explanation before they are forgotten. I would ask it as a favor of the citizens of Cerro Gordo to call on me to explain the mystery. C. GIRL.

I asked to be heard in council meetings, and I plead with them for a committee to investigate the church trouble, and see who was wrong, and if it was me I was ready to make a good confession; but no, the old bell-shepherd was the cause of all the trouble. He said to his flock, we must put this man down by denying him power, and we must expell him from our church in order to disarm him of all vestige of power that he might possess, and then we can go on and persecute him at will, and we can bring members of our church against him to establish whatever we may say against him. And that is the way they have done. They have resisted my rising in the world, and proclaimed that I am standing in the way of God and all good people. I have reached out for a helping hand to both great and small, but none responded and all were silent, and then I would become despondent, and found that victory of battle is not in the multitude or a host of men; but strength comes from heaven, and then I would renew my strength and fight on and on to victory or death, for no man sins who will fight for Jesus' sake.

A LETTER FROM MY DAUGHTER TO HER UNCLE.

CERRO GORDO, ILL., April 1st, 1888.

DEAR UNCLE: I feel it my duty to drop you a few lines this pleasant Sabbath morning. Pa received a letter from you yesterday and I was reading it last evening, so I thought it no more than my duty to tell you how Pa has suffered. I know just how he has suffered by all his friends forsaking him and not thinking for a moment what we were doing. I will not leave myself out. I have mistreated him with the rest. But I was blinded by evil influence, but now I can see what we have been doing. O, how sorry I am that I did not live closer to God's word, for that is the right way. I hope you will stop and think what you write to Pa, for he has suffered so much. Now, dear uncle, I don't mean this all for his good only, but for your own eternal happiness. I will tell you this because I know it is true, just as sure as God's words are true. If we all would have taken Pa's good counsel we would have spared ourselves so much trouble. But instead of doing that, we did not take heed to his advice. When we get good counsel that comes from the Bible, then we ought to receive it, for the time will come when all would like to receive it, when it will be too late. O, how important it is to prepare ourselves for that long and sweet rest. Time spent with God is never lost, for without him we can do nothing as it ought to be done. Pa has almost completed his history, and I am interested in it. I think it will be good, and hope it will bring souls to Christ. We are once more on the old farm where we were raised, and we all appreciate it more than ever. It is through God's goodness and mercy that we are here. Pa is the best friend we have except God. Now, uncle, what I have written I have written for good. We all have a soul to save, and I think just as much of your soul as any one's. This may be hard for you to understand, but it is true. My desire is to give good counsel. The way to understand this is to take God's word and see what He says. Everything shall fade away but His word, and that will always stand. I have talked to my brothers and sisters but they seem to get offended. I do not talk to them to do them harm, but want to show them that they have been led wrong. All must come to what I have sooner or later. God says why not come now, for to-morrow it may be too late. If a person will let his heart be hardened, then he can never see God. Dear uncle, I can tell you with truth that I have respected Pa's good counsel, for it was out of the bible. I pray that God may forgive me for not treating him as I should. O, how he has prayed for his family, that they might see their evil ways and turn to God. Always be sure and get right with God, then you will make a safe voyage across the sea of life, and a sure landing on the evergreen shore. I hope you will not think hard of me for writing. I hope to do you good by telling you what I know. Please tell your children to write. I will close by wishing you peace and happiness through this sinful world.

From your neice,
IDA GIRL.

Three Words of Strength.

There are three lessons I would write,—
　　Three words, as with a burning pen,
In tracings of eternal light
　　Upon the hearts of men.

Have hope! Though clouds environ now,
　　And gladness hides her face in scorn;
Put thou the shadow from thy brow,
　　No night but hath its morn.

Have faith! Where'er thy bark is driven,—
　　The calms disport the tempest's mirth;
Know this,— God rules the host of heaven,
　　Th' inhabitants of the earth.

Have love! Not alone for one,
　　But man as man, thy brothers' call;
And scatter, like the evening sun,
　　Thy charities on all.

Thus grave these lessons on thy soul,—
　　Hope, Faith and Love,— and thou shalt find
Strength, when life's surges rudest roll;
　　Light when thou else wert blind.

CLIPPINGS.

[From the *Decatur Bulletin.*]

The trial of the indictment of Louis Beery should be the last act in the Christian Girl tragedy. The history of this case is a remarkable one in a christian and civilized community, and smacks strongly of the witch-burning fanaticism of the olden time

Christian Girl is an old citizen and a prosperous farmer of Macon County, and was a devout member of the Dunkard Church; but, unfortunately for him, he is a man who does his own thinking, and does not hesitate to discuss the theological tenets of the church, and the practice of the congregation with the ministers, and the marriage of cousins, and the uneducated ministry were matters that Mr. Girl pressed with great vigor. Those of the brethren who were married to cousins, and the ministers who were unable to combat the logic of the heretic, declared him to be insane, withdrew his guarantee of salvation, and subjected him to the pains and penalties of a present and eternal boycott. He was driven from his own home and denied intercouse with the members of his own family, and information of insanity was filed in the county court and supported with great zeal; but the court and jury sat down hard on the movement by a prompt judgment in his favor, and the circuit court confirmed his vindication by a judgment of $500 damages for malicious prosecution. The indictment and conviction of Beery, his son-in-law, is for a brutal assault on the old man when he attempted to visit his wife, who was detained at the house of the defendant against her will, as he thought, will settle this uncivil controversy. We understand that Mr. Girl will take his case to the church council as soon as the matter in the court is settled.

Girl's Memoirs.
[From a Decatur Paper.]

Early this morning the reporter crossed palms with Christian Girl, who is now devoting all his spare hours to writing a voluminous history of his eventful life, covering a detailed statement of all his trouble with the Dunkard brethren in the vicinity of Oakley, where the old gentleman has a farm valued at $20,000. Mr. Girl has come into prominence through insanity and damage cases in the local courts, and is still after his alleged persecutors. A jury failed to give him a verdict for damages at the recent trial, and he will appeal the case to the Appellate Court. He has a room at Henry McDermot's residence, where the Memoirs will be corrected and revised before being handed to the printer, to be published in book form.

[From the *Decatur Bulletin*.]

Court was convened at 11 o'clock yesterday forenoon. The case now on trial is that of the people with Louis Beery. The defendant is a son-in-law of the redoubtable Christian Girl, who is the chief prosecuting witness. Beery is charged with having assaulted Girl with intent to do him bodily harm. D. L. Bunn is assisting the State's Attorney in the prosecution, and Beery is being defended by Nelson & Harnsberger and J. A. Buckingham.

[From the *Decatur Bulletin*.]

"You say you hear that I am insane? No, my dear brother, I am not crazy. Would to Heaven I were, or dead, or anything to be free from this awful

trouble. My church, my loving wife and my dear, sweet children against me, in a way that seems impossible to ever live together again, is breaking my heart, and it may drive me crazy in time; but, my dear brother, believe me, my mind is not gone yet."   Christian Girl.
December 15, 1884.

MR. THOMAS.

I will now expose a few more of what the Scriptures would term evil doers, such as would betray their trust for a little worldly gain, such as would come to a persecuted man and let on that they were in great sympathy with him, and even offer a helping hand. I will name as one, guilty of such cowardly work, an old man by the name of Thomas. He was condemning those treacherous Dunkards, and speaking rather hard of the dishonesty of some of them, and proposed that I should give him a job of trimming my orchard, and that he would help and get my family reconciled. I gave him a job of work and he demanded his cash as soon as the work was done, and he got his pay.

I then called upon him to do what he promised to do. He was to help to induce the old lady and the youngest daughter to go with me to Ohio, and there spend the winter, in order to get them away from bad influences, and the old coon endorsed my plan, and he said it was the best thing I could do, and he would help me all he could to persuade them both to go. He went to them and he made the old lady believe that my farm was all under mortgage, that some of the Dunkards told him that I owed more than I was worth, and the old lady was so worked up about it that Mrs. Marker told me that Mrs. Girl was restless all night from the scare from such hellish lies he tried to make her believe.

That is not all that pious looking old coon done. I took him to Mrs. Wells', where the youngest daughter was, and he had a private conversation with her, with the understanding between him and me that he should persuade her to go with her father, in order to induce her mother to go, so as to get away from bad influences, as the old lady was rather weak-minded on account of having her dear family broken up, and the general confusion of the neighborhood. The old man reported that the daughter refused to go. The daughter told me afterwards that he told her she had better not go with us to Ohio, and so, of course, we did not go, and he was the cause of my not going, and also the cause of other serious troubles. It is said that he has joined the Dunkard's Church. We do not know whether he repented of his treacherous act of hypocrisy or not. I suppose they, of course, would not refuse to baptize such birds as him, but as birds of a feather flock together, we think they ought to give him a position in the official board of that cousining ring to help them run their gospel ship overboard. We hope he may become a christian, and we wish him God-speed, giving him this admonition for his good and for the cause of Christ, and that he may be a true and faithful worker therein.

MR. ANTRIM AND WIFE.

We will now speak of some who call themselves evangelists, of Mr. and Mrs. Antrim, living in Oakley, Illinois, belonging to what is called the United Brethren. She is the best public speaker of the two, and, we think, wears the pants. She seems to know a great deal of Scripture, but we doubt whether she observes the plain commands of our Saviour's teachings.

I got acquainted with Mr. Antrim at a Sunday-school, and he left an appointment for his wife to preach on the following evening. I told Mr. Antrim that I was glad of his acquaintance, and the following conversation took place. I told him that I was a terribly persecuted man at the hands of false brethren. He heard me patiently, and seemed to sympathize with me. My request was that I was to meet him and explain to him in what way I was so mysteriously in trouble, but I went home and wrote him and his wife a long letter, asking them to answer or come and see me, or give me an invitation to come and see them; but no letter came, nor did they come to see me.

In my first talk with him I mentioned that Simon Nickey and his wife were leading two of my boys wrong, namely, William and Franklin, the oldest and youngest, who were under the devilish influence of these two hypocrites. He told me that our son Franklin was a member of their church. I asked him if Simon Nickey and his wife belonged to the same church, and he said they did. I told him that boy could never be a christian with such a load of sin on his shoulders, but he was making a very good start to be a complete hypocrite.

He did not answer my letter, so I went to his house one evening. He invited me into the house and bid me be seated. I asked him whether he got my letter, and he said he did. He was willing to answer my questions, which were fair and reasonable, but as soon as we got a little interested in the conversation his wife came bouncing out of the kitchen, not like an angel, but like a mad hare, and she did not want to be introduced to the stranger by her husband, but she was very angry, and declared that they wanted nothing to do with the matter, and I looked every moment for her to open the door and tell me to leave. She said: "Mr. Girl, I just tell you that we will have nothing to do with your matter;" and repeated that several times, and she wanted to do all the talking. I told her to hold up a little, that I wanted to tell her some things. I told her that I did not come to their house to quarrel with them, but to reason and plead with them. Well, I got her quieted down for a moment. I told her that I had a copy of the letter I wrote to them, that I was going to lecture at Sangamon on the following Sunday, and if they did not treat me right I would read it there. This made her more angry than ever, and told me to read it, she did not care, and I did read it. I will try and give the essence of it.

"CERRO GORDO, ILL., June 13, 1887.

"MR. ANTRIM: *Dear Brother*— I am sorry that I have not formed an acquaintance with you long ago. Now, dear brother, I have surely a work for you to do, and the best thing you can do is to say, Brother Girl, I will do what I can for you and your family. You know what is your christian duty, but, brother, I want a voluntary act, if any, with a christian spirit. Now, dear brother, here is a wonderful chance to do good, not only to your unworthy brother, but for the benefit of my dear inexperienced boys and my broken-up family. I feel that my soul is saved, but my children are in the hands of Satan's influence. You may say that Mr. Girl is one of those sanctified ones that cannot sin any more. No, no. I don't want to get close to one of those sanctified, pious-looking, religious fools. I want to be sure to take good care of my pocket-book as long as I am near them. He that says he has no sin is a liar. I would rather meet with a rattlesnake than one of those pious-looking preachers, for the snake will rattle before he bites, but a hypocrite will bite before he rattles. The old king said that I had to go to Jacksonville, but he is a liar. I am still here writing a history to expose his devilment. He has got the Devil's angels hovering round him in the very image of the Devil. Dear brother, don't get scared. It may seem to you that I am tearing something all to pieces. All that is good will hold until the end of time. He that is not for me is against me. We are to be hot or cold. Lukewarmness is condemned. And now, my good friends, you have a great work before you. Your wife is a colaborer with you in the spiritual work, and you ought to be a wonderful power for good; and as my dear and unexperienced son, Franklin Girl, belongs to your church, you may have a wonderful influence over him for good or evil.

"As I told you yesterday, that boy cannot be a christian until he releases his grip on his parents in holding secrets. Just think of it; he was influenced to swear falsely against his father. Don't try to reason with yourselves, and say Mr. Girl may be mistaken. No; I am not mistaken; but I am positive of the terrible thing,—of my boys whom I have raised, and God knows I have tried to raise them right, always admonishing them to be honest and to stick to the truth.

"Now, my friends, I will say you ought to be a happy pair, as your better half is watching over you as far as your spiritual affairs are concerned. The Scriptures say we shall be as wise as serpents and as harmless as doves. We will say that a great many of the preachers are as wise as serpents in accumulating worldly trash, as J. Metzker calls it. They have got that wisdom, but that is not the best part of the question. The latter part of that verse strikes the christian best. As harmless as doves. A wonderful contrast between the dove and the serpent. The one harmless and the other harmful as a hypocrite, and deadly and poisoning to a christian. The people are getting to throw a cloak of religion around them. You pull off that cloak, and behold, you have a ravenous wolf before you. They pray like a saint, but the Scriptures say, watch and pray. There is enough praying done to convert the whole world, and now, my friends, you have such before you in Simon Nickey and his wife; making pretentions that they are protecting one of those little lambs, who has sinned against his parents worse than the prodigal. Simon and his wife are leading him on in his evil ways."

Those two evangelists had a chance to do good, but they chose the evil part. I do not wish to screen one of them, for they are equally guilty with some of those Dunkards in helping to ruin and break up a once loving family. We hope those evangelists may learn a lesson from this book to do their duty as peacemakers, as any true evangelist should at all times be ready to speak peace to the church, to the community, and to the family.

We will give them a chapter to meditate upon, and hope they will be benefitted thereby. I Peter, chapter third: —

1. Likewise, ye wives, be in subjection to your own husbands; that, if any obey not the word, they also may without the word be won by the conversation of the wives;

17. For it is better, if the will of God be so, that ye suffer for well doing, than for evil doing.

also II Peter, chapter second: —

7. And delivered just Lot, vexed with the filthy conversation of the wicked:

8. (For that righteous man dwelling among them, in seeing and hearing, vexed his righteous soul from day to day with their unlawful deeds.)

We have strong reasons for believing that those two evangelists took an active part in hiding church members' ungodly mischief.

Wherefore, my dearly beloved, flee from idolatry.

I speak as to wise men; judge ye what I say.

For we being many are one bread, and one body: for we are all partakers of that one bread.

And now, my dear friends, as you cast me off as evangelists, not having as much as given me counsel, but treated me cold and indifferently I hope you will read my history carefully and prayerfully. I hope you will say to yourselves that you will do better next time; that you will judge for yourselves, and judge properly and truly and honestly, and pray to God to give you wisdom from on high, and try to obtain what you pray for. The weapons of our warfare are not carnal, but mighty, through God, to the pulling down of strongholds.

Now, as evangelists, it is your duty to go and preach and practice the bible in every land, and in every clime, and when a poor mortal comes to you for the bread of life, do not turn him from your door because he don't just suit you, or his multitude of sins are too great for you to battle with. May God help you to always hold up for the right, and do your duty to all mankind.

EXHORTATION.

Dear readers: I wish to bring the mysterious and strange conduct of my antagonists before you in as plain a manner as possible, often using our conversation in detail, so that all may understand it, and it is not an easy task for me to expose these evil men and women, but, in an illiterate way, I will give nothing but facts, and none of them can deny the truths I have written. I am just giving the actions of the leaders. If I should give you all of their hellish persecutions it would make a volume of a thousand pages, but I only want to give enough to the public to show what a set of evil men and women can do in breaking up a family and about break the hearts of the persecuted. They have caused me to spend hundreds of dollars in protecting my character. All good thinking people will see what I have had to stand in battling against a set of "professing" christians. I do hope they will reform, as that is partly the object of the publication of this book, and throw off the cloak of unrighteousness and get righteousness in their hearts, and then, instead of trying to tear down and break up families, they will try to build up and talk peace instead of confusion.

We find many blessed words in the sacred Bible to meditate upon. Among others I find the following, and if my persecutors had only meditated upon them they would have saved many a heart ache, and many a dollar.

They are found in Paul's letter to the Galatians, sixth chapter:

1. Brethren, if a man be overtaken in a fault, ye which are spiritual, restore such an one in the spirit of meekness; considering thyself, lest thou also be tempted.

2. Bear ye one another's burdens, and so fulfill the law of Christ.

3. For if a man think himself to be something, when he is nothing, he deceiveth himself.

4. But let every man prove his own work, and then shall he have rejoicing in himself alone, and not in another.

5. For every man shall bear his own burden.

6. Let him that is taught in the word communicate unto him that teacheth in all good things.

7. Be not deceived; God is not mocked: for whatsoever a man soweth, that shall he also reap.

8. For he that soweth to his flesh shall of the flesh reap corruption; but he that soweth to the Spirit shall of the Spirit reap life everlasting.

9. And let us not be weary in well doing: for in due season we shall reap, if we faint not.

10. As we have therefore opportunity, let us do good unto all men, especially unto them who are of the household of faith.

If they had only studied this passage of God's word they could have saved so much trouble. The first verse, if they claimed I was weak, it told them to restore me. The second verse tells us to bear one other's burdens, help to bear one another's grief, and make the load lighter upon a weak brother. The third verse instructs us not to think too highly of ourselves, but to get down humbly with our brother or we will fall; and in the fifth verse we find that every man must answer for his own sins.

I am glad that we do not have to answer for another's, for if I had to answer for some of these men's, I would be sure of being lost.

The sixth verse tells us that we must teach one another. When I began to teach them they were like a balky horse when the whip is used. It will kick up and try to get its harness off, and]when it starts up it will jump and do all it can to try and do you harm. Just as soon as I read the Word of God to them, and showed them they were not pulling they way God wanted them to, they began to kick, but they have not got the harness on to kick off. Not that I am a judge, only God says you shall know them by their fruits; and they are now trying to run away with my character, but I have a tight rein on them, and I think this book is the last turn I will have to make to bring them up all safe and sound.

The seventh verse tells us to be not deceived. God is not mocked. There is a great many of them who are trying to deceive the people by professing to be christians, but they cannot deceive God; and the latter clause tells us that whatsoever we sow that shall we also reap. I would not like to reap after some of them if they do not repent of their wickedness, for, if God's Word is true, it will be a very warm harvest.

Dear friends, I beseech of you to study these passages and follow them in the future, and you will reap life everlasting.

And now the tenth verse, there is a wonderful lesson in it: "As we have therefore opportunity let us do good unto all men." You all had a chance to do good, but you did not obey the teachings of that verse, but you did all the harm you could. Let us put on the whole armor of God, that we may be able to stand against all evil doings, and work boldly for our Lord and Saviour Jesus Christ, and he will reward us in the end.

ELDER STOUFER.

I wish now to bring before you, my readers, the actions of one more of the turn coats, and show how one more of my friends became a tool for that cousining ring after my trouble came up. Manon Stoufer was a little Dutch preacher who lived a neighbor to me for many years, and we were always on friendly terms before this trouble began, and after becoming my neighbor we talked very freely of church matters, and he agreed with me concerning the bad elements and rulings of church government. He was very free to condemn this marrying of cousins, and stated that he would not perform the marriage ceremony for them, and we were fast friends. When he moved away from our neighborhood to Millmine district he was elected to the highest degree of the ministry, and is now a bishop at Champaign county, Illinois, and after this he has always given me the cold shoulder and became very impudent to me on all occasions. He would come down to Cerro Gordo to partake with those evil doers in their pretended love feasts, and took their counsel in my case, and sided in with them, and obeyed the leader, Johnny Metzker, in all he said, and would not listen to what I would say to him, but he yielded to their wishes to help them cover their deception.

His colaborer in Champaign, Brother Barnhart, would come with him to attend their love feast, I think, by an especial invitation from old Johnny Metzker, for he was very anxious to make a good display around the communion table, and he did always manage to have a good supply of preachers to come to their feasts to deceive the outside world. The reader will see that this little Dutch preacher, M. Stoufer, has become a tool for those that he had poked fun at, and had called ignorant, and a selfish cousining ring governed by a little leader, one that is called Johnny. It is well enough to show to the reader what confidence Stoufer had in me. He signed a note as security for about eight hundred dollars. Those evil men got him alarmed. At least he said that I must stop traveling around on the cars or he did not feel safe on that note. I told him that was my business; if I wanted to travel that I was not under his obligation. I told him I would give him a chattel mortgage to secure him. I gave him to understand that he could not be my guardian, just because he signed his name on a note with me. I secured him after it was done, and the note was paid as soon as it was due.

The little Dunkard preacher had full confidence in me until he became a tool for them that he condemned, at one time, just as much as I did. About that time he became a middle man between me and John Mishler, to hold and deliver the papers between me and him on some borrowed money. Stoufer was to receive the money and send it to Mishler. The papers were sent to Stoufer, and he came to me and said: "You go down to Decatur and send that money to Mishler. I know you are honest, you can do that as good as I can; you are used to such business." I did so, and Mishler got his money, and sent me a good recommend like this :

"*Brother Girl:* Concerning our dealings for many years, I am glad to say that I have found you straight in all our dealings together, and must say that I always found you honest, honorable and set you down as a straight-forward business man."

But as Stoufer became a tool for those he had poked fun at, he has to suffer with them in having a place in this history, as he had plenty of warning with the many good and wholesome and truthful letters he received; but he was not willing to give a helping hand. He would not as much as answer one of my letters, but stood on the wrong side of the fence and grinned at me. My wife had confidence in Stoufer as a good counselor, and begged me to go to him for counsel. I went with her, and made them a visit, and Stoufer agreed to meet us before Dr. Barnes, in Decatur, Ill., and he was to be the judge whether I was a sane man or not. At that time, if he had had a good honest heart, he would have told Mrs. Girl, "You have got confused in your mind by that cousining ring, and you are a little stubborn. You have a sane man. All there is the matter is, your husband is fighting the devil that has come in the church, and you help him fight;" but instead of that, he agreed to meet us in Dr. Barnes' office on an appointed day, to lay the matter before him. John Phillips, and a few more, were appointed to accompany us to the doctor, and I agreed to open the subject before him.

Well, we met at the appointed time, but the doctor was absent, and Stoufer made an excuse that he could not wait any longer for the doctor, and that he must and would go home, and off he started like an old hypocrite, sneaking away. He had not been gone but a few minutes and the doctor came in. He sent a man after Stoufer to fetch him; that we were now ready to go before the doctor, but he sneaked off like a prairie wolf with sheep's clothing on. The rest of us approached the doctor, and I said: "Doctor, here are some folks that have been intimating that I was insane, and I agreed to come before you and have you to judge the matter." The doctor looked me in the face smiling, and said, "You seem to be all right to-day, but you might seem all right to-day, and not so to-morrow." I said, "Well, doctor, I guess I'll have to come and stay with you a week and saw wood for you." He laughed and we went home and attended to our business. Afterwards I begged Stoufer to go and talk to my confused family. He had promised to do so, but shrank from his promise, and no doubt but he was willing to help to confuse my family instead of talking peace to them. He was one of the strife-makers, and supported them in communing with them, and he was as impudent as an old rat, and as witty as a monkey. He was not brought to the court against me. It is a wonder that he was not, but he did enough mischief outside of the court-room. He tried to save others and can't save himself. He is only a middle-aged man, about forty five years of age. This will last him to the end of his career, and we hope he may come out and make an open confession to my wife and children for neglect of his duty as a shepherd, and be a peace-maker the balance of his lifetime, and do the Lord's will, and not side in with a set of ungodly men to stand as an evil-doers and encourage obstinancy in a man's family. Blessed are the peace-makers, for they shall see God.

JOHN METZKER.

Dear Reader: I will now endeavor to speak of the leader in all my trouble and the leader of the cousining ring. I have written much about his word is "law" in "his church." I say his church, because he paid most of the money to build it, and he uses that as a great authority, and says "my church." He forgets that it has been dedicated to God and His cause, but some have to be very important, and he is one of them. His actions show the cowardly, unchristianized principle that so many people have in heart, that when a man starts down hill, to push him all they can instead of obeying our Saviour's commandments,— to do good to all mankind. His money gains him many followers, and they all do his bidding. He uses his money like the Pharisee,— after giving it he makes a great blow about it. Our Savior tells us that our righteousness must exceed the righteousness of the Pharisee, or we cannot enter the kingdom of heaven; and again he tells us that we must not let our right hand know what our left hand doeth, but Johnny must surely spend about as much advertising as he gives. He was about the first man I became acquainted with in this place, and our first acquaintance was very agreeable to me. I found him very friendly, and he seemed to want to accommodate me. He heard that I wanted to purchase a farm, and he told me of different farms for sale, and I bought a quarter section in the same neighborhood he lived in, and he accommodated me in various ways. I thought I had found a lifetime friend, but I found him to be a Satan in disguise. He has loaned me money on my word, and through his many kind deeds toward me he gained my confidence, and I appreciated his kindness; but it seems to me that it was only a bait, and I have learned a lesson by it. Experience teaches us many a lesson, but we have to pay dear for them; and now, through his influence, I have had to undergo the vilest persecution that a person could stand and come out a sane man. I was invited many times to unite with his church, and by their good preaching my wife was induced to unite with them, and in a few days I followed her example. I started with the full determination to live a christian life and obey my Saviour's teachings, and I have done my best, but under the circumstances it has been almost impossible, but I will hold out faithful until the end.

After persuading me to join the church he was not satisfied until he had me out, and if a man with money wants to run a church or any thing else in this nineteenth century he can do it. Christ says in Matthew "that there is nothing impossible with God;" and "he" is a god of the Cerro Gordo brotherhood, and, of course, there is nothing impossible in his ruling the church; but I hope they will see that *that god* can not save them. If they follow in his steps he will lead them to destruction, if he does not turn in the right direction and repent of his evil deeds. May he remember the calling he has is the highest calling given to man, and I, with Paul, will say:

"I, therefore, the prisoner of the Lord, beseech you that ye walk worthy of the vocation wherewith ye are called.

"With all lowliness and meekness, with longsuffering, forbearing one another in love.

"Endeavoring to keep the unity of the Spirit in the bond of peace."

My prayer is, that he might read this book carefully and see the harm he has done; how he has been the leader in breaking up a once loving family, and causing discord in a whole neighborhood. Paul says: "Walk worthy of the vocation wherewith ye are called," and I hope he will try in the future to take Paul's advice in this respect, and practice what he preaches.

EXPENSE ACCOUNT.

This is a partial statement and list of the court and other expenses that this vile cousining ring has brought upon me. It is the remnant of an effect brought on by trying to live strictly honest and religiously among a set of hypocrites and liars.

Sheriff Fees,	$202.50
Witness Fees,	430.00
Lawyer Fees,	855.00
Doctor Fees,	90.00
Last time,	150.00
Rent,	57.00
Traveling,	80.00
Nurse Fees,	16.00
Total,	$1,880.50

It is impossible to make a strictly correct statement of all the cost those brethren have forced on me by their bad evidence, but this I can say, five years ago my farm was all clear of mortgages, and I had some money loaned out on interest, and had my farm well stocked with horses, cattle, hogs, sheep and poulty, and now I have my farm doubly mortgaged for about four thousand dollars. In the four years I lost four thousand dollars, to say nothing about the wear and tear of my old body, as Johnny Metzker would say. Better wear out than rust out. Wear the body out to save the soul. He said that right if he would only practice what he preaches, then he would be a good man. We will still hope he will repent and be converted to God. He has been converted and perverted, and is now confused in his old age. I hope he will muster up courage enough to come out on the Lord's side, and lay the devil in the shade. As he is so anxious to have a good name in the Dunkard's papers, we hope he will now come out through the *Messenger* with a good confession. I would rather go to heaven through the poor house, than go down to hell in a golden craft.

Paul says: "Charge them that are within this world, that they be not high-minded, nor trust in uncertain riches, but in the living God who giveth us richly all things to enjoy." I hope he will cease to be high-minded and get down on a level with his brethren who have not so much of this world's goods.

BROTHER MQORE.

A few lines for Brother Moore of Kenka, Florida. His treatment towards a persecuted brother is not to be recommended. I visited his place in a far away country with the expectation of having a warm reception from him and his little band of brethren. I went eighteen miles, from Palatka to Kenka, in order to see him and the brethren, to form some acquaintance with his little flock, which he had under his care, as he is their elder and has started up the enterprise by way of a church there to advance the cause of Christ. I supposed he was the main man in the enterprise, and I went there to worship with them, but got rather a cold reception from him. He did not treat me as though I was a persecuted one, but rather as though I was an impostor, while I yet belonged to the church before this great conflict commenced so rough. At that time Brother Moore was yet living in Champaign county, Illinois. I have entertained him and he has partaken of our hospitality. I told him in a letter that he owes nothing for that, but I don't know how it will stand between him and his God, to point out a boarding house to a brother that he knew for many years, who had come to see him and make him a friendly visit. As we stepped off the train he pointed me to a boarding house that was kept by some of the members of his church. That did not disappoint me, for I had money to pay my way. The next day I attended their meeting and shook hands with him, with some hope that he would give me an invitation to come and see him as his house was in plain sight. Well, I thought it was rather cold treatment, and on Monday I went back to Palatka, where I and daughter Ida wintered for a few weeks. I took my daughter Ida up to Kenka, as she had some lady friends that she had known in Illinois, by the name of Wolf. They entertained us pleasantly, and it was very much appreciated. God will bless them for their kindness towards some of their fellow men.

What I have to say about them is not to flatter them. I hope it may be my privilege to entertain them, and show my respect to them for their kindness shown in time of need. We went to meeting together, and Moore did not as much as invite me and my daughter to see them, and did not as much as introduce his wife to us. It made my daughter spunky. She did not care about going to hear Brother Moore preach. He showed not a very good spirit just at that time, but I hope he will repent of his cold treatment to his Illinois friends. It hurt my daughter's feelings worse than what it did mine. She became a member of the Brethren Church, and I think she made a good start in the divine life. It is possible that we will go to Florida again, and I have wondered whether Brother Moore will treat us any better if he is convinced that I was a persecuted one instead of an impostor. He said in the first and last letter that he did not think he had treated me cold. I could say a good deal more, but a hint to the wise is sufficient. We will make some allowance for Brother Moore. The trouble is he places too much confidence in what I call big guns while he preaches in the pulpit,— that God is no respector of persons. The bishop is getting too high-minded. I will refer Brother Moore to the twenty-third chapter of Luke. Bro. " More " has forgotten more Scripture than I ever knew. I hope for the time when Brother Moore will say that he has learned some things from one who has been called a fool, or crank, by some of his colaborers whom he put so much trust in.

LAST LAW SUIT.

I will now give some of my experience of the last law suit in what is called a court of justice, the law suit brought before 'Squire Yoder, on the 15th day of February, 1888. I shall first name the cause of the law suit between me and my tenant, William Heil. He was introduced to me by J. M. Rainey of Decatur, Ill., and was recommended as a good and honorable man. Mr. Rainey said his wife was partly raised in his family, and he spoke highly of her and said she was as neat as a pin. There may be pins that were lost, and they may become rusty, so they become useless. Well, we will see whether they filled the bill according to Mr. Rainey's recommendation. J. M. Rainey acted as a guardian over Heil. As he had taken an active part in the law suit before 'Squire Yoder, we think that he done the planning in the suit, although he seemed to act very innocent in testifying about the amount of apples in the orchard, also the quality of the apples. He made them so very inferior that they were hardly worth noticing. He said they were little bits of snarly apples hardly fit to use. The lawyer asked him if they were as large as grapes. I think Rainey said, there were only a few snarly apples. He said he was not over all the orchard, and that he did not notice, and could not tell much about the quantity or the quality. The lawyer asked if Heil brought him some of the apples, and he said he believed he did bring some to his house. When asked how many, he guessed about a peck or so. When asked if he ate some of them, he said he did not remember whether he did or not, and yet he seemed to know that they were hardly fit to eat, — all wormy and no good. He stated that I called on him to speak to Heil in order to quiet him down so that a man could talk and reason with him. He did not tell that I said, that if he did not quiet down I would have a peace warrant served upon him, that Heil had made serious threats, etc. Rainey said he had talked to him, but could do nothing with him. He acted as guardian over him, but by his conduct and his evidence, displayed a willingness to protect him in his devilment; as much as to say — " I can do nothing against him, but I will do all I can for him, even if I do have to strain the truth."

Rainey, by his very actions, showed that he knew he was not doing right in supporting such a man, who he had recommended to me. He was ashamed to own that he acted as guardian over Heil. Heil asked his consent to rent the summer-house. Rainey took a very active part in helping Heil to cheat me out of my apples. He was willing to call them little bits of snarly things, althouh he had sold some of them as high as one dollar a bushel. If Mr. Rainey had given good counsel to Heil, Heil would have done better. It seems to me that I never saw a more guilty-looking set before a 'Squire in my life, and I think J. M. Rainey was their bell-sheep. He makes no profession of religion. I think he told me he used to belong to the Episcopalians, but he either left the church, or they threw him overboard. I had lots of dealings with the man. I bought some cattle from him, paid him some money down, and gave my note for the balance. I soon found out his shrewdness. That was the first of my acquaintance with Mr. Rainey. I was forced to sell the cattle sooner than I wanted to, because my antagonist got a fee-bill against me of $640, and my tenant, Heil, was contrary, and would not let me have my own corn to feed my cattle with, so we shipped the cattle to Chicago in Rainey's name, and I went along to see to them. After we received remittances from our agent, I paid off some lawyer fees and other expenses, and had enough left to pay Rainey. He said he did not want the money, he only wanted interest. I told him I could not use the money to good advantage. He had found out by that time that my land was heavily encumbered, but he was willing to take a second mortgage to secure his $840. I suppose he calculated to get my homestead of eighty acres, for he thought it had to be sold to pay court and other expenses that a set of devilish liars had forced on to me.

To prove my assertions, I will give a further account of the proceedings of the old shrewd fox's plans to get what the lawyers and the hypocrites did not get. It seems that the cunning Heil was to be Rainey's stool-pigeon to catch a good thing for the old backslider. After Heil had been refused the farm for another year he was then ready to condemn the land as poor thin soil, and wanted to know whether I would take fifty dollars an acre for it. I said no, and then he offered fifty-five dollars, when I told him that he could tell Rainey that the least money that would buy it was eighty dollars per acre. He said Rainey would never give that. They then tried something else on the supposed crank. They knew that I was cornered up close by those breachy Dunkards, so I suppose they had it all fixed up in their own minds that they would have a good haul on the cranky fellow. Heil did the fishing for Rainey, cut the bait, etc., and they fished and fished, but they could not catch the crazy fellow. They did not have the right kind of bait. Heil got angry because he did not get the farm for another year, and Rainey put out more bait. Heil carried the news back and forth. Rainey sent word the second time that I should come to town; that he had an eighty acres of land down in Christian county, five and a half miles from Niantic, that he wanted to show me, that he wanted to trade me for my eighty. I went to Decatur, and he hired a team from the livery stable and took me down. I looked at his river-bottom land. It was a very dry time, and he explained how easy it could be tiled, and then it would be such a valuable piece of land. There was no fence on it, a little shell of a house on a knoll at one end of it, which was the only dry spot on the whole eighty acres. There was a pond of four or five acres, which I believe he called a fish pond, and he said there was no trouble to tile it off. He showed me a ditch somewhere about the center of the eighty, and I asked him what that ditch meant, and he said he supposed it was to run the tiles into, but I found out it was merely a levee to prevent the water from going over the whole farm. Mr. Howard told me that tiling would not do any good in that bottom; that the whole bottom was overflowed about every year, just long enough to ruin their crops, and his wife said that people who had land in that bottom were worse off than if they had no land at all, on account of the overflows. But Rainey tried to make me believe that a little tiling was all that was necessary to make the land very valuable. He was very much afraid I would inform myself about the land in that vicinity. As it was about noon, his tenant wanted him to put the horses in the stable, and have dinner, but he said he would drive somewhere in the shade and feed, and eat a lunch.

I made some inquiry about my old friend, Mr. Steward, who lived in that vicinity. He did not want me to hunt up my old friends, because he knew they would post me about that river land. We drove about two miles, unhitched, and ate a lunch. Rainey told me to get my book and pencil, "and let us figure." I did so in order to see what he would do. I made up mind that I did not want his land. I figured my land at seventy dollars per acre, and held his at forty dollars, but he said I asked too much for mine. I told him I knew he asked too much for his frog ponds. I then figured, and it seems the difference was something like $400, and he wanted to know whether I would split the difference. I said, "No." "Well," said he, "you study over it," and we then hitched up and drove home. The next day I took the train and went to Illiopolis; within three and a half miles of the land we went to see the day before, and met my friend, Mr. Samuel Howard. I went home with him, and he went with me to see the land, and he told me it overflowed every year, and that it might be worth about ten dollors per acre. I came home and he, Rainey, dropped me the following:

"DECATUR, ILL., July 14, 1887.
"MR. C. GIRL, Oakley, Ill.,

"*Dear Sir:* I have been waiting for you in reference to our trade. If we don't succeed in making the trade I will ask you for what money you are owing me, as I wish to invest elsewhere. Please let me hear from you at your earliest convenience, and oblige,

"J. M. RAINEY."

I answered his letter and told him that his land did not suit me, and that I would not trade, but I would sell him my farm at $85 per acre, and that was lower than I ever offered it to any one else. What he wanted was to skin me out of it. Mr. Rainey's card:

"MR. GIRL: Your postal is at hand. Whether we make a trade or not I do not care to discuss the matter. You no doubt recolleet that we agreed on figuring, and you insisted that I should go and see your land, which I agreed to do on Saturday. According to the understanding, I went, and you, instead of being there, I learned, was running over the country getting advice from others. I was careful not to misrepresent the land to you, which you admitted I had not. You said further that you expected you would have a big fuss about signing the deeds. I told you rather than that should happen to let the trade go. You said no."

I will show that I have a very fair memory for an old man, and I do not remember any such language to have taken place. My dealings here with the citizens of Macon and Piatt counties for about twenty-four years show a better record than what Mr. Rainey can show to the people. As I have said, he is a shrewd trader, and, I think, a skinner. He says, in another letter, "I certainly considered it a trade, and was very much disappointed, not so much by not making the trade, as I was in the treatment you gave me, in not coming near to explain yourself."

I am now explaining myself, and Mr. Rainey had a timely warning by a postal card, but since he came to assist his dishonest friend Heil, he showed his colors on the witness stand, but tried hard to hide his deception. They had a timely warning not to join in with such dishonest people as Heil and his wife. Here is the card I sent him:

DECATUR, Ill., Nov. 27, 1887.

"MR. RAINEY: If you desire to avoid exposure of your sly acts in trying to play a smart game on me in a land trade that you had proposed to me, you should have said that you made a fair and square trade with me, which I say is not true."

I have written a second card to him, and the reader will see for himself what Rainey was driving at. He was watching to get what those breachy Dunkards and my tenants could not get, and I will name the lawyer, too, who was trying to help them. He is one of the kind the Rev. Sam Jones refers to, that could be tolled to hell with nickels.

We will now give the card to the public to read. I don't know what possessed him to send those cards back to me unless it was for me to publish them. It seems to me that I would have burned them.

DECATUR, Ill., Dec. 6, 1887.

"You have no use for such cards, and you have sent it back to me. Much obliged. Please send the one back about the land trade. I say it is a lie and here is the proof; 'if we don't succeed in making a trade, I will ask you for the money you owe me.' The real truth is you did not want your money, but wanted to force a trade on me. I will say again I am much obliged to you. I am not afraid to redeem my paper."

Lawyer Buckingham thought that he was going to prove that they had just cause to arrest me as an insane man, but he got beat at his own game. He must acknowledge that he got hold of the wrong Girl, and a Christian at that. A Christian Girl is not to be trifled with. A hypocrite cannot down a Christian, and a backslider is not much better than a hypocrite.

Rainey said in a letter:

"I regret very much that you and Heil cannot get along. It was through me that he concluded to let you in the house. I think he feels disappointed in not getting the place for another year. You had him believe there would be no trouble about it. I think the best thing you could do would be to buy him out.

J. M. RAINEY."

I tried to buy him out. I offered him more than his crop was worth,— more than I could have made out of it, and out of those rotten apples. If I had bought him out I would have had some apples to eat this spring. I will say that Rainey took a very active part with that prairie wolf, because I would not let him cheat me out of my farm. He saw that I would not let him have his way, so he turned in and helped Heil to play the devil.

When I first commenced dealing with Rainey I thought he was reasonably fair, but he turned out like the man's mule. "A mule is good to you all week to get a chance to kick you on Sunday." An old man like Rainey ought to be competent to give a young man good counsel instead of supporting him in such ungodly things as he did. I hope when they both read this history that they will repent of their hellish work, and ask their Creator to forgive them for their sinful conspiracy in playing a sharp game on a man that has been persecuted almost like our Saviour. They must know they have a soul to save or lose, and I think a man as old as Rainey to engage in such a hellish work as he did ought to be cut off from society. I would not carry such a guilty conscience as he showed before 'Squire Yoder for all the property he owns, for it will not buy his way into eternity. I hope that he may realize that an honest man is the noblest thing of God's creation. Just think, Mr. Rainey, we must come face to face with our God, and then we will say O, God. Then them rotten apples will stare you in the face.

JACOB MILLER.

I will now proceed to speak of one of the "old order" elders. He was one of our brethren. He turned "old order," and became a great worker with them. He said my antagonists were too much for me, and would not stay with them and try to get them right. He left the ship and jumped on to the flatboat, or jumped out of the frying-pan into the fire. May God help such apparent saints, who wear all their christianity on the outside and none on the inside. But let us see what David says in the fifty-first Psalm, sixth verse :

"Behold Thou desireth truth in the inward parts, and in the hidden part Thou shalt make me to know wisdom."

They are upholding members in the church who went into court and testified unto untruths. Such government in a church is worse than none. Paul says in I Corinthians, fifth chapter, last verse :

"But them that are without God judgeth. Therefore put away from yourselves that wicked person."

We know that liars and backbiters and all such are ungodly men and women, but they still keep them in their church; but these old evangelists, who have been preaching so long, do not take our Lord's and Apostles' doctrine, "to speak boldly the words of God," for they are afraid they will hurt some one's feelings, and it would be money out of their pockets. We find that the Word of God is sharper than a two-edged sword, but the old preachers do not cut anyone with it. I hope they will see where they are wrong, and bring those who have sworn to lies before them, for they will always be stumbling-blocks for the outsider to fall over. Let us see what the Bible says in regard to this, and I pray you will learn this chapter by heart.

Turn to that book of praise and prayer, the first Psalm:

1. Blessed is the man that walketh not in the counsel of the ungodly, nor standeth in the way of sinners, nor sitteth in the seat of the scornful.

2. But his delight is in the law of the Lord; and in his law doth he meditate day and night.

3. And he shall be like a tree planted by the rivers of water, that bringeth forth his fruit in his season; his leaf also shall not wither; and whatsoever he doeth shall prosper.

4. The ungodly are not so: but are like the chaff which the wind driveth away.

5. Therefore the ungodly shall not stand in the judgment, nor sinners in the congregation of the righteous.

6. For the Lord knoweth the way of the righteous: but the way of the ungodly shall perish.

We see by the foregoing the happiness of the godly, and the unhappiness of the ungodly.

This old elder knew how they were ruling their church, and he left them to better himself, but he is upholding the same ruling in the church he is now in. He said he was too weak to overcome them then, and I don't think he has become any stronger yet; but God says, come boldly to a throne of grace that you may find help in time of need, but you cannot rebuke them with God's word, for you might lose a member and a dollar. We can hardly tell what christianity is by looking at the lives of so-called christians in some localities, where they are ruled by the devil's agents, and men that say they are too weak and sickly in God's cause to fight for the right. Jesus did not say "I am too weak," when he was with the devil, but He said, "Get thee behind me, Satan."

And now let me say to all these ungodly men, I will not have you in our church and allow you to walk with me, but get behind me devils' imps, or repent of your wickedness and do right.

This old elder is very weak in the teachings of our Lord and Master, but it pleased him to see that I was too much for the hypocrites. He said that he was not strong enough to meet such trials as I did, and said it would have been better if there had not been any division in the church, and said those bishops stood by each other in all things. Some people will be too weak to climb up to heaven, if they are too weak here on earth to confess Christ. They cannot get the power after they leave here; and because he is weak he will go hand in hand with false brethren, and has not courage enough to get them out of their ungodly ways. He is sustaining one that came to court and testified falsely, and is upholding him in this offence, that is against the laws of our country and our God. He told the old elder that he swore just as he believed, and they are willing to keep him with them. I hope that this short exhortation will teach the old elder a lesson, and may he grow in strength and knowledge of our Lord and Master.

ETERNITY IS DRAWING NIGH.

Pray, brethren, pray. The sands are falling,
Pray, brethren, pray. God's voice is calling.
    Yon turret strikes the dying chime,
    We kneel upon the edge of time.
  Eternity is drawing nigh!

Praise, brethren, praise. The skies are rending;
Praise, brethren, praise. The fight is ending.
    Behold! the glory draweth near,
    The king himself will soon appear.
  Eternity is drawing nigh!

Watch, brethren, watch. The day is dying;
Watch, brethren, watch. The time is flying;
    Watch as men watch the parting breath,
    Watch as men watch for life or death,
  Eternity is drawing nigh!

Look, brethren, look. The day is breaking,
Hark, brethren, hark. The dead are waking,
    With girded loins already stand,
    Behold! the bridegroom is at hand.
  Eternity is drawing nigh!

EXHORTATION.

Dear reader, it is a very difficult task for an unlearned person to write a history, and especially all they lack is to take me out and kill me. In all these hours God has not forsaken me, but has kept me through it all, and I, with Jesus, will say, forgive them; lay it not to their charge. I have often thanked God for those blessed words, found in Paul's second letter to the church at Corinth, twelfth chapter, eighth and ninth verses:

"For this thing I besought the Lord thrice, that it might depart from me.

"And he said unto me, my grace is sufficient for thee: for my strength is made perfect in weakness. Most gladly, therefore, will I rather glory in my infirmities, that the power of Christ may rest upon me."

If it were not for those blessed words we would often yield to temptation. "My grace is sufficient for thee." But I have always found God's promises to be true. He will fulfill all his promises if we will do our part. In our most trying hour, when all our friends forsake us, we find, by putting our trust in him, that his grace is sufficient for every trial, and in all temptations he will make a way for our escape.

Peter, who was always the spokesman for the Apostles, steps to the front and says: " The Lord knoweth how to deliver the godly out of temptation." I have proven this promise to be true, for God has known how to deliver me, and he has not allowed me to be tempted above that I was able.

Dear reader, I hope you may learn a lesson from this feeble exhortation, for we are all to be tried, and God has given us all a way of escape if we will only use it. Let us put on the whole armor of God, and let us wield at all times the sword of the Spirit, which, as Paul tell us, is the Word of God.

May God help all to be faithful to the end, and assist all hypocrites to repent of their evil ways, is my prayer.

A HINT TO ALL.

Beloved Bishop and Elder: You are getting things off in your favor. You say in your reply to my request that I said you were no christian man or you would have answered my letters, and Mr. Deck told you what was not so. You say such are about the exact words used, as parties in the house at that time can testify. We may testify to anything, and the judges do not know but what we are telling the truth; but oh! who can judge human hearts? I am ready to give an answer of the blessed hope that is anchoring my poor soul. Human aid seems to be out of my reach, so I will do as the poet says:

> "Cast all your cares on Jesus,
> And don't forget to pray."

I will have a good victory if all those preachers shoot at me with all their wisdom, and all their shrewdness, to try and hide their deception. They may hide their deceptions from poor mortal man and woman, but what do they preach to their fellow man, that God is an allwise being, knowing even the intents of the human heart. Have you the spirit of Christ with you, to treat a poor persecuted man as you have treated me? Did you not say, the time I approached you, that I have a legal right to call on my fellow man for aid when I am in distress? Did you act out your part as a christian, when I approached you with tears in my eyes? Did you set me down as a Dunkard impostor, or a crank, or a religious fool? Well, my good friend, I am willing to be called a fool for Christ's sake, but I do not like the idea of being called crazy by a set of money-gods and skeptic preachers. God knows there are hundreds of them filling the pulpits, making pretensions that they are leading their congregations right, but are leading them to ruin and destruction. There are thousands of poor souls led astray. Can the blind lead the blind; shall they not both fall in the ditch?

I would rather see the devil come at me in the form of a demon, than to see a skeptic put into the pulpit to preach. Great God! where are our American people drifting to? Can you answer the question? It seems to me that they are drifting into idolatry as fast as the wheels of time can roll them along, and still some of our well-educated preachers are crying peace, peace, when there is no peace, but strife and confusion. They that forsake the law praise the wicked. How much better are the poor, who walk in their uprightness, than he that is perverted in his ways though he be rich. He that turns away his ear from hearing the law, even his prayer shall be an abomination unto God.

EXHORTATION.

The title of this book will show to the world what it contains, and I hope it will be studied with care, for I write it for the benefit of all christians and sinners; to show to the christian that there is power in the love of Christ; and for the sinner, to show what a christian can stand, and exhort them to become christians. This volume is also for the benefit of hypocrites, to try and get them to turn from their evil ways, and stop making a mockery of God's cause. I will not leave out the skeptics,— for they are always ready to stand up and pray, and make a mocking of our blessed Bible. I have had all these to handle, and I found it very hard, but a true christian can handle them with "God's Word" at all times. I started out to be a Christian, and I will be, let it be life or death ; and now, after all my persecutions, I will pray with the psalmist David:

1. O Lord my God, in thee do I put my trust; save me from all them that persecute me, and deliver me:

2. Lest he tear my soul like a lion, rending it in pieces, while there is none to deliver.

3. O Lord my God, if I have done this; if there be iniquity in my hands;

4. If I have rewarded evil unto him that was at peace with me; (yea, I have delivered him that without cause is my enemy:)

5. Let the enemy persecute my soul, and take it; yea, let him tread down my life upon the earth, and lay mine honor in dust. Selah.

8. The Lord shall judge people; judge me, O Lord, according to my righteousness, and according to mine integrity that is in me.

9. Oh let the wickedness of the wicked come to an end; but establish the just: for the righteous God trieth the hearts and reins.

God will deliver us from our enemies if we put our trust in Him. I know there is a vast number of good people in the Dunkard Church, but I have been persecuted by a set of devils in that church, and by some of other denominations. We can read that the saints will judge the world,— it it is not necessary for me to name them in this, but I have named them in the foregoing chapters, at least some of them, and they can all see what they have done. May God help them all.

PRAYER.

O Lord God, creator of all things, who art fearful and strong and righteous and merciful, the only gracious King, the only giver of all things, the only just, almighty and everlasting King; thou that delivered us from the ungodly and from all trouble, please, our Father in heaven, gather together those who are scattered from us; deliver them that serve Satan; look upon those that are distressed and oppressed; let the heathen know that thou art our God. If not against thy will, punish those that oppress us and with pride do us wrong, and plant thy people again in thy holy place. Be merciful to those who have no mercy on themselves. My defence is in thy power and in our dear Saviour, thy Son, who was sent upon the earth to die for us. Thou hast promised, through thy dear Son, that the believer in Christ shall not be persecuted above that he is able to bear. We come, O, our Heavenly Father, to thee for help; make us strong in thy word and thy power. Come then, thou blessed Lamb of God, deliver us out of the hands of our persecutors, make us strong in our faith, and help us to provide all things honest before God and our fellow-man, so that when we are done upon this sinful earth we may have an assurance of meeting beyond this vale of tears, where we will be forever blessed. Kind and merciful God help us to this end, and thine shall be the praise, honor and glory forever more. AMEN.

www.ingramcontent.com/pod-product-compliance
Lightning Source LLC
Chambersburg PA
CBHW031333230426
43670CB00006B/338